X-ray Radiation and Artificial Bragg Structures

SCIENCES

Waves, Field Director – Frédérique de Fornel

X Ray, Subject Heads – Philippe Jonnard and Chouki Zerrouki

X-ray Radiation and Artificial Bragg Structures

Coordinated by
Jean-Michel André
Philippe Jonnard

iSTE

WILEY

First published 2025 in Great Britain and the United States by ISTE Ltd and John Wiley & Sons, Inc.

ISTE Ltd
27-37 St George's Road
London SW19 4EU
UK

www.iste.co.uk

John Wiley & Sons, Inc.
111 River Street
Hoboken, NJ 07030
USA

www.wiley.com

Library of Congress Control Number: 2024947082

British Library Cataloguing-in-Publication Data
A CIP record for this book is available from the British Library
ISBN 978-1-78945-187-0

ERC code:
PE2 Fundamental Constituents of Matter
PE2_12 Optics, non-linear optics and nano-optics

Contents

1

Historical Prolegomena

Jean-Michel ANDRÉ and Philippe JONNARD
LCPMR, CNRS, Sorbonne Université, Paris, France

1.1. A short history of sources of X-rays

As the story goes, on November 8, 1895, Wilhelm Conrad Röntgen stayed on at his laboratory late into the evening because he was intrigued by the observations he had just made. He was working on what was known as cathode rays using a Hittorf discharge tube. He was attempting to create total darkness in the room in order to detect the visible radiation (fluorescence), emitted by his device, using a fluorescent screen of barium platinocyanide. To do so, he covered the tube with a black envelope, opaque to visible light. When he turned the tube on, he saw a glow that disappeared when he turned it off. He noticed that this glow came from the fluorescent screen. Surprised by this, he sought to understand fully, as a thorough experimentalist would do, and continued his investigations: the intensity of light increased when he moved the screen closer to the tube. He continued his work after dinner and noted that what seemed to come from the tube was able to pass through thick screens of paper or wood (several centimeters thick) but could be stopped by lead.

He worked hard for several weeks to try to find out what these "rays" were (*X-Strahlen* as he called them) while keeping his discovery secret. Shortly before Christmas, he informed his wife Bertha and took a radiograph of her hand. On December 28th, he sent a report on his investigations to the Würzburg Medical Physics Society, whose Chairman agreed to publish it, even though no oral presentation was made at the time. Röntgen presented his results on January 23, 1896 to a meeting of the Physics Society where he explained:

X-ray Radiation and Artificial Bragg Structures,
coordinated by Jean-Michel ANDRÉ and Philippe JONNARD.

> By chance, I found that the rays passed through black paper [...]. I then used the instrument as a camera, and the experiment worked.

At the end of this meeting, Röntgen successfully made a radiograph of the hand of the anatomist Albert von Kölliker, who proposed to name these rays X- or Röntgen rays, and this suggestion was approved. However, it is of note that the scientific publication presenting the discovery of X-rays does not mention the fortuitous nature of his discovery.

The "official" discovery of X-rays is a good example of a type of serendipity: it is highly probable that the production of X-rays was observed prior to Röntgen's discovery, but that the observers were not truly aware that they had discovered a new type of ray. Sir William Crookes himself, the inventor of the eponymous X-ray tube, observed irradiated photographical plates that had been placed near his tube, as did the doctor Arthur W. Goospeed. The physicist Yvan Pulyui, using a tube of his own design, noted photographical plates that were also irradiated by the radiation from his tube. However, Röntgen, with his expertise as an experimentalist, is responsible for the scientific approach (verification, attempt as characterization, etc.) that led to what can be called a scientific discovery. Serendipity played a full role in this.

Very quickly after Röntgen's discovery, many scientists set to work to reproduce Röntgen's experiments and determine the characteristics of this radiation. In particular, we can cite the work of Jean Perrin who repeated several of Röntgen's experiments and also demonstrated that X-rays were capable of discharging charged electrical bodies and of ionizing (in modern language) gases and metals (the "metal effect", using his expression) (Perrin 1948, 1956). Léon Gouy, a close friend of Perrin, established that if these rays behaved as waves, their wavelength must be considerably shorter than 1/100 of the wavelength of green light. It was not until Max von Laue's work on crystalline diffraction was completed in 1912 that the wave-like nature of these X-rays was demonstrated definitively. The "tormented" history of the discovery of the nature of X-rays, their physical origins and the crystalline diffraction is described in detail in the published work of Authier (2013).

In terms of research, the schools of Cambridge and Edinburgh became the main centers for X-ray spectroscopy, with in particular the works of Henry Moseley who successfully recorded the emission spectra of a large proportion of the elements in the periodic table, before his tragic death in 1915. His works contributed considerably to the construction of the atomic model by Niels Bohr. A long period of interaction ensued between newly discovered mechanics and the experimental results in the field of X-ray. Therefore, the multiplets observed by Alfred Landé in the X-spectra of heavy atoms led Wolfgang Pauli and Ralph de Krönig to the electron spin hypothesis.

Concerning the production of X-rays, improvements were made progressively. Until the 1920s, the main source remained the Crookes tube, which is based on a discharge in a gaseous environment under a partial vacuum which deteriorates over time (pressure ranging between 7 10^{-4} and 5 10^{-5} Torr). The electrons that generate X-rays come from ionization of the residual gas molecules, and are excited by the action of continuous electrical voltage from a few kV to several hundreds of kV. This voltage is then generally generated by Rühmkorff coils. Tubes based on this mechanism are sometimes referred to as cold cathode tubes. In fact, the Crookes tube is not entirely suitable for producing X-rays because it has several disadvantages for this task:

– The vacuum is "imperfect", which damages propagation of the electrons that collide with the gas molecules.

– To produce hard X-rays, the electrical voltage between the cathodes must be increased, which creates electrical beakdown issues.

To mitigate these problems, the solution consisted of creating a high vacuum (pressure lower than 10^{-6} Torr) within an airtight chamber, generally a glass bulb sealed after the pumping to create the vacuum. The electrons are produced by a thermo-ionic effect: a cathode that has been heated, generally directly by the Joule effect, emits, due to this thermal agitation, electrons that are then accelerated by high voltage (from a few kV to several hundreds of kV), as seen in Figure 1.1.

X-rays are produced when the atoms of a target known as an anode or sometimes anticathode[1] are bombarded with the electron beam, by means of two effects: *Bremsstrahlung* or braking radiation, and characteristic emission[2]. The first effect is due to deceleration of the electrons by the nuclei of the target's atoms. The intensity of the phenomena increases as the atomic number of these atoms increases, which explains why the anode is composed of heavy atoms, generally tungsten. The second effect is quantic in nature: it corresponds to the emission of energy in the form of electromagnetic radiation when an electron jumps from an electron shell to fill an electron hole; in the case of the X-ray tube, the ionizations are caused by the

1. The Crookes tube has three electrodes: a cathode and an anode between which the HV is applied, and a third electrode connected to the anode and positioned at an angle of approximately 45° to it. In the Coolidge tube, there are only two electrodes: an emissive cathode and a second (anode), sometimes also known as an anticathode, which is the target for "projectile" electrons and emits X-rays.

2. The term X fluorescence also exists but the term is not correct because, in this case, the ionization must result from an irradiation by X-rays and not by electrons.

electrons, projectiles of the cathode beam. William D. Coolidge, who worked at General Electric, is responsible for fine-tuning the design of the eponymous tube in 1913 (see Figure 1.1), implementing the aforementioned principles. This type of tube was made in different forms in accordance with the works of several "inventors", generally doctors specializing in radiology. Even though it is more effective in terms of X-ray production than the Crookes tube, the efficiency of the Coolidge tube is only of the order of 1%. However, it remains the most frequently used source of radiation: classic radiology, scanners, tomography, spectroscopic analysis, etc. It is, or has been, made by several companies: General Electric (GE), Compagnie générale de radiologie (CGR), Philips, etc.

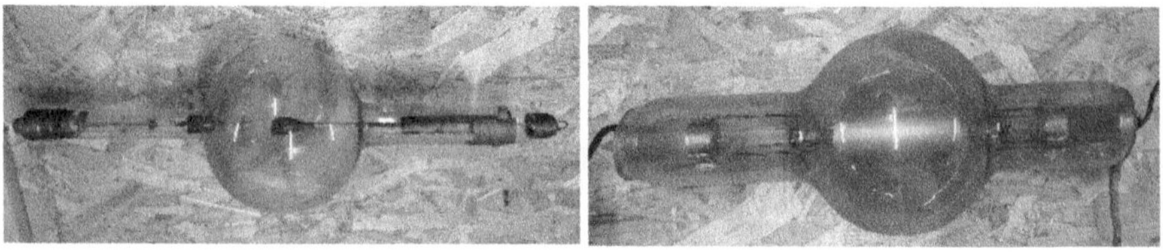

a) b)

Figure 1.1. *a) Coolidge X tube; b) kenotron (© P. Jonnard)*

We will mention that, while the X-ray tube itself was being manufactured, tubes known as kenotrons were also manufactured to rectify the alternating voltage and obtain the direct voltage required to create the HV or VHV field that generates acceleration of the projectile electrons. The kenotron is in fact a diode (valve) that can be mounted on a bridge rectifier to ensure full-wave rectification. Manufacturing of these requires the use of technologies close to the X-ray tube technologies themselves: vacuum technologies, glasswork, etc.

For specific applications requiring very hard X-rays, in other words with a photon energy nearing MeV, the source of "projectile" electrons can be made up of an electron accelerator, generally known as a LINAC (linear accelerator), or a "betatron" circular accelerator. Use of a Van de Graaff or the Cockroft–Walton electrostatic generator remains anecdotal.

In the field of research, the situation changed in the 1960s. Everything began with the observation of radiation known as synchrotron radiation. It is electromagnetic radiation with a spectrum ranging in practice from infrared to hard

X-ray[3], emitted by charged particles (electrons, or positrons) accelerated to a speed close to the speed of light, when they are subject to a centripetal force resulting from the action of a magnetic field. This phenomenon occurs in accelerators known as synchrotrons and storage rings. Initially, it can be considered a parasite effect for this type of machine, since it causes a loss of energy of orbiting particles that must be compensated for (Iwanenko and Pomeranchuk 1944). This radiation was also named orbital radiation.

Although the concept of synchrotron emission appears in various scientific works at the end of the 19th century (Liénard 1898), observation of it dates back to the 1940s: indirectly by the variation of particle orbits in 1946 by John Blewett and directly in 1947 by the observation of visible light on a 70 MeV synchrotron by a GE team. Driven by eminent scientists including W. Coolidge, the father of the X-ray tube, different circular accelerators were constructed in the United States, including the 70 MeV synchrotron at GE's laboratory in Schenectady. One of the problems occurring with this type of machine was intermittent sparking. Attempts were made to locate this visually in the machine using a mirror through a port that opened onto the interior of the machine. During an inspection of this type, on April 24th, 1947, a technician known as Floyd Haber saw a spot of light coming from one of the tubes. Having already eliminated the Cherenkov effect as the cause of this emission, physicists including Robert Langmuir and Herbert Pollock came to the conclusion that it was Schwinger radiation[4] (Pollock 1983). The circumstances of this observation caused a certain degree of controversy, and the pre-eminence of the discovery was contested on several occasions, in particular by George C. Baldwin (Baldwin and Kerst 1975). Be that as it may, this synchrotron radiation was considered by the "machine engineers" either as a problem, or as a way of characterizing their apparatus.

It was not until the 1960s–1970s that physicists specializing in X-ray spectroscopy became interested in synchrotron radiation as a potential source of X-rays. Professor Yvette Cauchois (see Figure 1.2), who at the time was director of the Chemistry Physics Laboratory (LCP) at the University of Paris, and a group of Italian scientists including Ugo Fano played a key role[5] in this area. Fano, who worked with Enrico Fermi and discovered the eponymous effect, had to immigrate to the United States

3. The spectrum range depends on the energy of the accelerated particles. The higher this energy, the more "hard" components it presents, in other words photons with higher energy levels (ultraviolet, x-ray).
4. Julian Schwinger, an American theoretician and physicist, performed a theoretical study of synchrotron radiation, first in 1946 then in 1949 (Schwinger 1949).
5. From 1927 onwards, Fermi set up a group of talented theoreticians including Ugo Fano, Ettore Majorano and Gulio Racca who focused on the application of the new quantum mechanics with regard to X-ray spectroscopy.

where he ended up at the National Office of Standards, the forefather of the prestigious National Institute of Standards and Technology (NIST).

Figure 1.2. *Yvette Cauchois in front of an X-ray machine (©Laboratoire de Chimie Physique - Matière et Rayonnement)*

Cauchois, aware of the interest of synchrotron radiation for X spectroscopy, met Fano who shared the same point of view. A fruitful working partnership ensued, named *Sanita Luce*, between the French and Italian teams following communications between Cauchois and Mario Ageno, who was at the time director of the physics laboratories of the Italian National Health Institute. The Italians had already worked on characterization of the X-ray emitted by the ADONE storage ring in Frascati, whereas the LCP had great expertise in X-ray spectrometers. The particle physics communities, Frascati's Italian one and the French at Orsay, had previously worked together, in particular during the transfer of the Italian ADA accelerator to LINAC in Orsay. Cauchois created a team of young experimentalist researchers, Pierre Jaeglé,

Pierre Dhez, Christiane Bonnelle, Robert Barchewitz, and technicians, including Henri Ostrowiecki; among the Italians, there were G. Missoni and G. Cremonese. Concerning hard X-rays, the LCP had a Cauchois spectrometer, operating by transmission, and a Johann spectrometer, operating by reflection under vacuum. Bonnelle was responsible for these crystal spectrometers; concerning soft radiation, a grating spectrometer had been developed by Jaeglé and Dhez in the new section of the LCP installed at Orsay. This group of scientists and technicians made up the pioneering team in X spectroscopy using synchrotron radiation.

It appears that the first spectrum obtained was an aluminum transmission spectrum in the vicinity of the K absorption threshold (see Figure 1.3). The conditions of this first result were described by Ostrowiecki (2014). The priority sometimes claimed for this spectrum obtained using synchrotron radiation may be debated insofar as D.H. Tomboulian and P.L. Hartman, in the mid-1950s, characterized the radiation emitted by the 300 MeV Cornell synchrotron using a grating spectrograph and the Be K and Al L thresholds (in the field known as Vacuum UltraViolet (VUV) and not in the field of X-ray) had been observed (Tomboulian and Hartman 1956). Be that as it may, following this initial success obtained from the French–Italian collaboration, a series of publications on the results of X-ray spectroscopy (Cauchois et al. 1963; Jaeglé and Missoni 1966; Jaeglé et al. 1967) and X-ray optics (Barchewitz et al. 1967, 1969) ensued. The works of Jaeglé and Dhez in Frascati were in the so-called Holweck field, to observe the emission lines of strongly ionized atoms in order to detect possible population inversions that would potentially lead to a laser effect in this domain. This turned out to be the case, and Jaeglé is considered the father of X-ray lasers obtained using hot plasma.

After a change of direction by the Italians, the working relationship ceased in 1971. The Italians recycled the Frascati synchrotron into an optimized source of synchrotron radiation and continued their works in the field of XUV and of soft X radiation (10–200 eV), with in particular Antonio Biancini in the Soliditi Roma group of spectroscopists. In France, Cauchois contacted scientific decision-makers to promote synchrotron radiation. In 1963, she contacted André Blanc-Lapierre to obtain assurance from the Laboratoire de l'Accélérateur Linéaire (LAL), which was in charge of constructing the ACO collider ring for particle physics studies, that a magnet with a strong magnetic field would be included for the purpose of producing hard X-rays; the forefather of the wiggler. This vision, far in advance of its time, unfortunately did not come to fruition for technical reasons and due to issues about sharing the machine with particle physicists, it would appear.

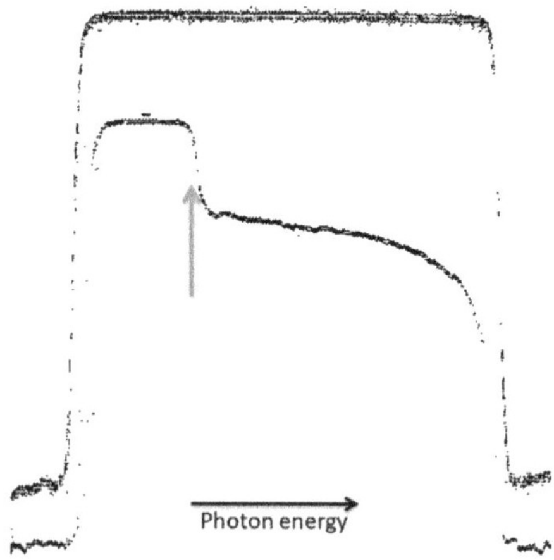

Figure 1.3. *Aluminum transmission spectrum obtained on the Frascati synchrotron on May 6, 1963 by a French–Italian team. The upper curve shows the continuous base obtained through a beryllium film; the lower curve shows the spectrum obtained through an aluminum screen. The green arrow indicates the discontinuity in K absorption of the aluminum. Figure adapted from Cauchois et al. (1963)*

In 1972, Cauchois brought the subject up again, now equipped with the results achieved at Frascati. Thanks to a well-constructed line project known as VUV, the cluster of young scientists (Yves Farge), including those who had returned from Frascati (Jaeglé, Dhez, etc.), and the support of influential scientists at the CNRS (Jacques Friedel and Hubert Curien), success was achieved! The considerable role played by the LCP in the use of synchrotron radiation in spectroscopy and X-optics is analyzed in the article by Bonnelle and Dhez (2014), and the history of this joint French–Italian effort is described in the article by Mottana and Marcelli (2013).

A laboratory called Laboratoire pour l'Utilisation du Rayonnement Electromagnétique (LURE), provided with funding and personnel, was created in 1971; in the spring of 1973, the line of light from ACO's magnet 8 was opened. Several lines of light, to employ the terminology used in synchrotrons, were then opened on the ACO, allowing original research on many themes. The physics and chemistry of atoms, molecules and of solids held an important position: atomic spectroscopy (photo-absorption) in the VUV field with, among others, Pierre Jaeglé, Pierre Dhez and François Wuilleumier, supported by theoreticians such as Michel Cukier and Françoise Combet-Farnoux (Cukier et al. 1974); X-ray photoelectron spectroscopy led by Jean Lecante's team; angle-resolved photoemission where Yves Petroff (who was director of the LURE then of the ESRF) and his team played a key role. This research led to many firsts (Thiry et al. 1979; Petroff and Thiry 1980):

– first complete band structure of a metal;

– illustration of the effects on N-bodies (in nickel);

– first observation of surface states on core states.

In addition, research in the field of X-ray optics using synchrotron radiation, which began with the working partnership between Robert Barchewitz and the Italian team, and was at the origin of firsts in what later became known as ReflXANES and ReflEXAFS (Barchewitz et al. 1978), continued with Jean-Michel André; this gave rise to the first determination of the continuous refraction index (real *and* imaginary parts) in anomalous diffusion regions in the field of X-rays (André et al. 1982). Experiments in the anomalous zones were then highly successful, in particular involving magnetic multilayers.

In 1975, ACO was abandoned by particle physicists and dedicated completely to the use of synchrotron radiation. The other collider on the Orsay campus, Dispositif de Collisions dans l'IGLOO[6] (DCI), opened up to the use of synchrotron radiation in June 1976 with cohabitation of the communities: particle physicists and synchrotron radiation users. The X-rays at this facility are harder than those available at ACO.

Research using synchrotron radiation was launched with first-generation equipment in various countries (Tantalus I in the United States, DESY in Hamburg, etc.). These would rapidly find themselves limited, and the race was on to create second-generation machines designed entirely to optimize the production of synchrotron radiation, obtain strong brilliance[7] and make researchers' work

6. IGLOO is the name given to the building that resembles, with its rounded shape, an igloo made from ice; it initially housed the targets room of the Laboratoire de l'Accélérateur Linéaire.

7. Brillance is a quantity that qualifies the source; it is expressed in photons/s/mm^2/mrad^2/0.1% BW, where BW refers to the bandwidth.

"comfortable"[8]. With the experiments carried out on Tantalus I at the University of Wisconsin in the United States, the superiority of the electron storage rings as a source of synchrotron radiation was demonstrated. In a storage ring, the electron beam circulates continuously for hours, whereas in synchrotron machines, repeated injections are required. Sources of synchrotron radiation using storage rings were then constructed in the most industrialized countries: National Synchrotron Light Source (NSLS) at Brookhaven in the United States, Adalin at the University of Wisconsin (United States) succeeded Tantalus I, ALS at Berkeley (United States), the Photon Factory at the KEK laboratory in Japan, BESSY I in Berlin (Germany), Doris II and III in Hamburg (Germany), SSR in Daresbury (United Kingdom), Super-ACO in Orsay (France), MAX-I in Lund (Sweden) and Adone in Frascati (Italy).

Nevertheless, a whole swath of research was not yet accessible with this type of second-generation machine because the X-rays delivered were not hard enough, and due to insufficient brilliance. This is the case in particular for radio crystallography, with its significant application in the structural determination of molecules of biological interest. It became necessary to create so-called third-generation machines offering the required characteristics. Yet, the cost of these machines was becoming very high because delivering photons with greater energy requires larger machines and the particles in orbit in the machine have higher energy, as shown in Figure 1.4 showing the universal curve of the spectrum of synchrotron radiation produced by a curve magnet. It is characterized by the critical wavelength $\lambda_c(nm) = \frac{1.864}{B\,(E_p)^2}$, with B (in Tesla) being the magnetic field of the magnet and E_p (in GeV) being the energy of the particles in orbit in the machine. The smaller the value of λ_c, the more the spectrum tends towards hard radiation.

In addition, relatively costly so-called insertion elements (wiggler, undulator) must be included to increase the efficiency of radiation production. These elements installed on the straight sections of the machines are assemblies of N_d magnetic dipoles alternated on a regular basis, which impose a quasi-sinusoidal trajectory on the electrons. They are characterized by a parameter, K, which takes into account the period of the assembly and the intensity of the magnetic field: if K is less than 1, the insertion element is an undulator (K is typically 0.5) and the intensity emitted varies as a function of $N_d{}^2$ following an interference phenomenon between the emissions at each dipole. On the contrary, if K is greater than 1, the element is a *wiggler* (K is typically 3) with emitted intensity varying as a function of N_d.

8. Working conditions for the first devices were sometimes difficult; in the ACO bunker, it was very hot in summer and difficult to stand up straight, with access only by ladder to top it all off!

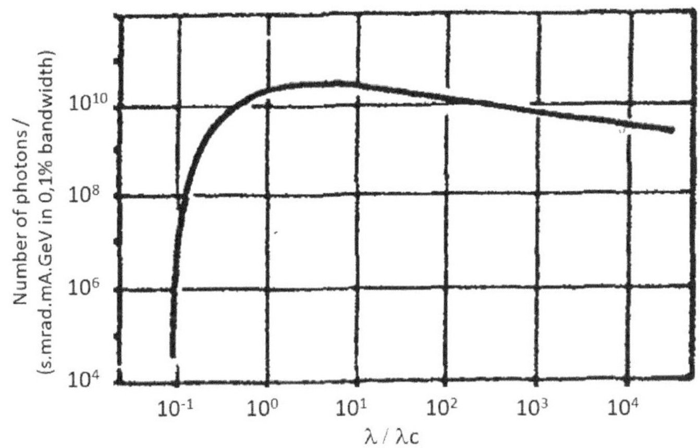

Figure 1.4. *Universal spectrum of synchrotron radiation for 1 GeV electrons as a function of the length of the scaled wavelength* $\frac{\lambda}{\lambda_c}$ *(λ_c is the critical wavelength). According to Bessière (1996)*[9]

The third-generation machines delivering soft X-rays were created at the national level (SOLEIL in France, Diamond in the United Kingdom, BESSY II, Petra III and ANKA in Germany, Elettra in Italy, SLS in Switzerland, Pohang Light Source in Korea, etc.) but for high energy machines (energy of an upper beam of 5 GeV) delivering very hard X radiation, collaboration between different European nations was required to build the European synchrotron ESRF, or investment by large countries capable of funding very large instruments (VLIs). This is the case for the Argonne National Laboratory (APS) in Chicago in the United States, PETRA at DESY in Hamburg and Spring-8 in Japan. Continuous improvements were made to most of these machines by introducing different techniques: continuous injection, slicing, etc.

In the field of X-ray production from an electron beam, an important step was made in the works of John Madey. In fact, in the 1950s, H. Motz at Oxford (United Kingdom) became interested in what is known as undulator radiation and proposed to use this device as a "light amplifier". His work was followed up by GE engineers who developed the *Ubitron* (Phillips 1960), consisting of a progressive wave tube where an interaction is established between an electron beam undulating under the

9. See doi: 10.1051/jp4:1996449.

effect of a magnetic field and an electrical transverse mode from a waveguide. We can refer to the forefather of the undulator used in storage rings to produce X-rays. These works were classified and "buried" until they were "exhumed" by J. Madey at the beginning of the 1970s. He succeeded, in collaboration with the High Energy Physics Laboratory team at the University of Stanford (United States) (Elias et al. 1976), in constructing a device based on the effect then known as the stimulated *Bremsstrahlung* (Madey 1971) that he named the Free Electron Laser (FEL).

Overall, the first FEL consisted of an undulator associated with an optical cavity used to confine the radiation within the undulator. The electrons were initially injected using a LINAC. However, Y. Petroff, having returned from the United States where he took part in a conference held by Madey, busied himself with installing an FEL device on the Super-ACO storage ring. Therefore, the first FEL operating on a ring was created; it was named the Centre pour Laser Infra-rouge d'Orsay (CLIO) because it emitted within the infrared spectrum by means of a mirror cavity. The presence of an optical cavity of this kind then restricted the FEL to the production of long wavelength radiation. A phenomenon known as self-amplified spontaneous emission (SASE) would then lead to the avoidance of use of the cavity and thus allow the wavelength to be shortened (Bonifacio et al. 1984). The SASE phenomenon is based on the use of noise (instability of the electrons) in a high gain nonlinear regime, causing saturation in a single pass, provided that the undulator is of sufficient length (Huang and Kim 2007). This operating mode, named SASE-FEL, is independent of the wavelength but given that it starts with a noise, it remains chaotic in nature, which manifests itself in instability on a spectral and temporal level and in the intensity of the radiation emitted. Figure 1.5 shows the principle of SASE-FEL.

The radiation is emitted in the form of a series of pulses with an envelope that has a temporal width of the order of tens to a few hundred femtoseconds and a spectral width of the order of tens of eV. Other than its "extraordinary" brilliance peak (10^{32} photons/s/mm^2/mrad^2/0.1% BW), this SASE-FEL radiation presents a transverse coherence due to the gain-guiding effect that synchrotron radiation does not offer. In order to improve the temporal coherence and reduce the bandwidth, different techniques have been proposed and tested or are currently being tested. This is mainly the seeding technique. In the field of extreme UV (EUV), seeding was performed using high-harmonic generation (HHG) emitted by a gas pumped by an infrared laser (Togashi et al. 2011). A wide variety of external input techniques have been developed: HGHG, EEHG, PEHG (Huang et al. 2021).

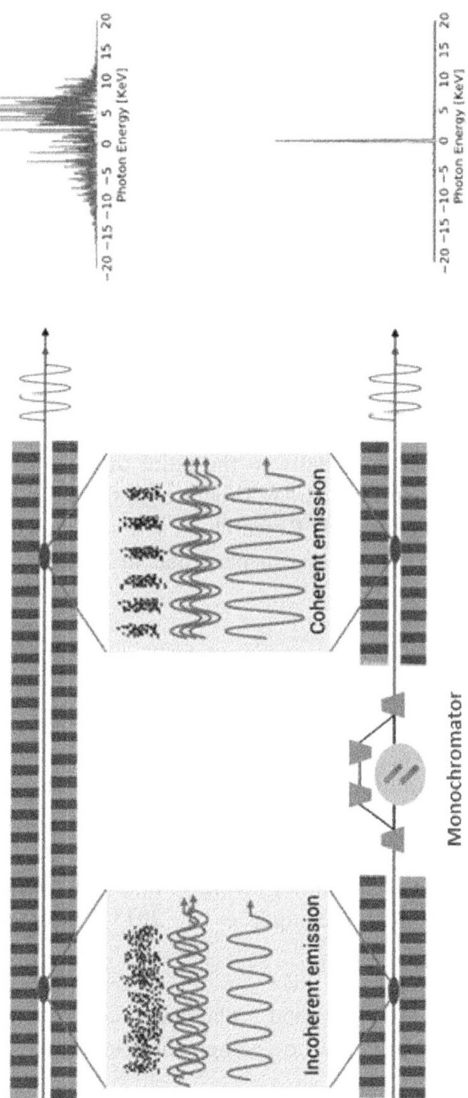

Figure 1.5. *Two operating modes of an FEL*

COMMENT ON FIGURE 1.5.– *a) SASE mode; b) self-seeding. At the entrance to the undulator (modulator), a spontaneous emission of radiation occurs in a linear regime with low gain; the radiation interacts little by little with the electrons which are in micro-packets and then emit coherently; the radiation follows a nonlinear regime with amplification and lastly saturation. In self-seeding mode, a monochromator is placed prior to amplification, which occurs in the radiator, to "purify" the radiation. Figure adapted from Huang et al. (2021)*[10].

In the field of soft and hard X-rays, another technique known as self-seeding, which does not use an external source of radiation, was invented (Feldhaus et al. 1997). This consists of reinjecting filtered or monochromatized radiation into the SASE-FEL radiation using a Bragg monochromator (hard X-rays) or a diffraction grating (soft X-rays). Figure 1.5(b) shows the principle: (at least) two undulators are used. The first constitutes a so-called "modulator", which operates in SASE mode and in low gain regime, whereas the second, sometimes named a radiator, operates under a high gain regime, potentially to the point of saturation. The technique of self-seeding could benefit artificial Bragg structures in the XUV and soft X-ray fields, as we will demonstrate in Chapter 8. Research on self-seeding continues to be highly active in the various XFEL centers, in particular at the European XFEL (Geloni et al. 2012).

The performances of FEL in terms of brilliance (high intensity), temporal structure and coherence have already resulted in a multitude of results; among these, we can cite in particular:

– for brilliance: illustration of the self-amplified spontaneous emission and stimulated emission in the field of hard X-rays (Yoneda et al. 2015), soft X-rays (Rohringer et al. 2012; Beye et al. 2013) and EUV (Jonnard et al. 2017), intensity-induced X-ray transparency (Young et al. 2010; Stöhr and Scherz 2015), stimulated Raman emission (Weninger et al. 2013), saturable and two-photon absorption (Hoffmann et al. 2022);

– for the temporal structure and coherence: coherent diffractive imaging (Chapman et al. 2006), Fourier transform holography (Gorkhover et al. 2018).

Table 1.1 lists the main XFEL installations around the world, along with their characteristics. Figure 1.6 shows the historical evolution of sources of X-ray based on the use of electrons (or positrons) and of their brilliance. Eduard Prat's recent article

10. See doi: 10.1016/j.xinn.2021.100097; from the journal *The Innovation*, a Cell Press partner journal.

(Prat 2021) and the references included in it provide a more in-depth discussion of the principles of synchrotron radiation sources and free electron lasers.

Installation	Energy of the beam (GeV)	Energy of the photons (eV)	Rate of repetition (Hz)	Width at half-maximum of the pulse length (fs)
FLASH	0.35–1.25	14–620	4,000–10^6	10–200
LCLS	2.5–16.9	2,800–12,800	120	5–400
SACLA	5.1–8.5	4,000–20,000	60	2–10
FERMI	1–1.5	20–310	50	30–100
PAL-XFEL	3.5–10	275–20,000	60	5–100
SwissFEL	2.1–5.8	250–1,240	100	1–20
European XFEL	8.5–17.5	240–25,000	2.7×10^4	3–150
SXFEL	1.0–1.6	124–1,000	50	30–1,000
LCLC-II	4–15	200–25,000	120–10^6	1–500
SHINE	8	400–25,000	10^6	3–600

Table 1.1. *Performance data for the main XFEL facilities around the world*

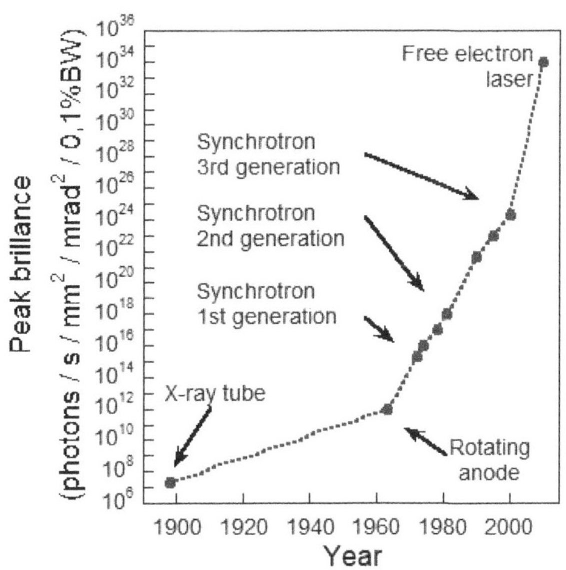

Figure 1.6. *Changes in sources of X-rays over time, based on the use of electrons (or positrons) and their brilliance*

Concerning the production of X-rays using relativistic electron beams, other techniques based on different phenomena can be implemented. We can cite inverse Compton scattering, channeling radiation, transition radiation and parametric radiation (Rulhusen et al. 1998). The latter two can benefit from artificial structures, as we will see in Chapter 8. However, we point out that to date, none of these techniques has resulted in facilities comparable to the synchrotron which delivers radiation on request for users, but that various projects for compact sources are in progress or being finalized. Therefore, we will mention the ThomX project in Orsay (France), targeting the production of radiation at 45 keV by Compton scattering (Bruni et al. 2016), the parametric X-ray (PXR) source project in Japan (Hyun et al. 2018) and X-ray source projects using channeling (Piot et al. 2012).

Aside from the production of X-rays from an electron beam, another technique has been put to good use: emission by highly ionized atoms generally obtained from hot plasmas[11]. An ionized atom is likely to emit characteristic radiation. The advantage of hot plasmas is that under certain conditions, it is possible to obtain a population inversion, as P. Jaeglé had predicted despite the criticism that he faced[12], with, as a consequence, the possibility of stimulated emission and ultimately laser emission. Jaeglé's group played a considerable role in the development of this type of laser, and currently a laboratory resulting from the works of his group, named LASERIX, and which is a platform of the University of Paris-Saclay (France), provides laser radiation in the XUV field; this platform offers the scientific and industrial community access to a wide range of coherent sources of radiation. Several other facilities operate around the world: the Compact Multipulse Terawatt (COMET) system at the Livermore National Laboratory (LLNL) in the United States, Prague Asterix Laser System (PALS) in Prague (Czech Republic). Development of this type of XUV laser has benefitted from multilayer artificial structures for the creation of the optical cavities that provide the retroaction. The groups led by Matthews et al. (1985) and Suckewer et al. (1985) played a pioneering role in this field.

Interaction of the laser beam with matter is also a source of short wavelength radiation. The characteristic of the radiation depends on the characteristics of the laser (power, spectral field, etc.). For medium power lasers emitting femtosecond pulses in the infrared spectrum, the phenomenon of HHG allows radiation to be produced with a spectrum up to the XUV domain. Laser radiation is focused on a jet of rare gas which uses a nonlinear jet to generate attosecond pulse trains (Drescher et al. 2001). This type of quasi-coherent source is frequently used in experimental studies of ultrarapid phenomena in atomic and molecular physics (Catoire et al. 2014). These sources benefit from Bragg multilayer optics.

11. A plasma is a state of matter composed of ions and electrons from ionized atoms.
12. The probability of stimulated emission decreases as wavelength decreases.

Interaction of an ultra-intense TW (10^{12} W) or PW (10^{15} W) laser beam operating in a femtosecond regime with solid matter leads to the emission of radiation with very short wavelengths up to gamma rays. The electric field of the incident laser causes oscillation of the surface of the target at the laser's frequency with speeds ranging from zero to the speed of light, forming a so-called oscillating mirror. At its heart, the electrons vibrate at frequencies ranging from the laser's fundamental frequencies up to high harmonics reaching the X-ray domain. The Apollon laser, installed at the Saclay plateau (France), will allow this type of operation to take place and will soon be followed by extreme light infrastructure (ELI) European facilities (Amiranoff 2016).

These various radiation sources have allowed exploration of a large part of the spectral domain of X-rays. Depending on the physical phenomena studied or on the many technical means implemented to disperse and detect the radiation, the X-ray domain has been subdivided into several sub-domains (see Figure 1.7). The limits separating these various domains, which can be further divided, are arbitrary.

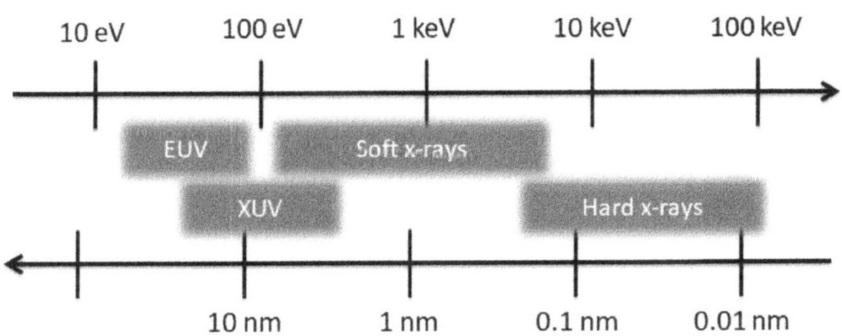

Figure 1.7. *The electromagnetic spectrum, with X-rays in the high energy or short wavelength domains, divided in turn into various sub-domains; EUV is the extreme ultraviolet domain*

1.2. A short history of artificial Bragg structures and changes in them

1.2.1. *Older work*

This book is about artificial Bragg structures and their application to X-rays. This type of structure resembles a photonic crystal, as it is now agreed that it should be called. Structures that we will be considering have periodic modulation of the optical index according to a spatial dimension and will therefore be unidimensional, corresponding to a one-dimensional photonic crystal (1D-PC).

The term photonic crystal, introduced in 1989 by Yablonovitch and Gmitter (1989), has its origins in the close analogy between the propagation of electrons in a crystal and the propagation of the photons in a (generally artificial) periodic optical structure. The terminology for photonic crystals has been analyzed by Yablonovitch in an article in which he insists on the need for a significant contrast in the optical index in order to characterize a photonic crystal. As for us, we will use the term "photonic crystal" for a periodically modulated optical environment in the spatial dimension and which creates strong coupling through this periodic arrangement with the electromagnetic radiation that propagates in it, in this case X-rays. In practice, this implies that the period involved is close to the wavelength of the radiation or to a multiple of it.

Long before this concept of photonic crystals, important figures in physics tackled the issue of propagation of an electronic or electromagnetic wave in a periodic structure. Illustrious names in the field of mathematics and physics between the 17th and 19th centuries can be associated with the propagation of a wave (acoustic, optical, etc.) within a one-dimensional network. Among these, we can cite Isaac Newton for the problem of the speed of propagation of sound; Jean Bernoulli and Daniel Bernoulli jointly for the principle of superposition; Augustin Cauchy and William Thomson (Lord Kelvin) for the study of wave dispersion; Joseph Fourier for the problem of heat propagation. Historically, it appears that John W.S. Rayleigh, in his works in 1887–1888 (Rayleigh 1887, 1888), following those of George G. Stokes, can be considered responsible for the discovery of the notion of a band gap, associated with an increase in reflectance with acknowledgment that the problem is mathematically connected to George W. Hill's equation.

In 1946, the publication of the famous work by Léon Brillouin, entitled *Wave Propagation in Periodic Structures* (Brillouin 1946), set down the academic foundations of theoretical physics concerning wave propagation in material environments with a spatially periodic structure. In it, the author presents diffraction of X-rays by crystals and the dynamic theory pioneered by Max von Laue, Paul P. Ewald, Charles G. Darwin, Walther Kossel, William H. Bragg and William L. Bragg (father and son) to mention just a few. Their works constitute the basis of modern physics of photonic crystals under the meaning we attribute to it.

In 1928, Bloch (1929) expanded on the results of Floquet (1883) to draw up a theory of propagation of electrons in crystals, which would be transferred to the propagation of photons in photonic crystals.

1.2.2. *Modern work*

In the 1960s, theoretical works concerning periodic structures related mainly to the exact resolution of the wave equations in sinusoidal or laminar environments, to spatial–temporal propagation in periodic environments and to radiation from sources included in such environments. From the late 1970s onwards, considerable technological innovations occurred, coupled with significant progress in modeling: active materials were introduced in connection with retro-distributed lasers, anisotropic materials, nonlinearity, magneto-optical, etc. All these contributions continue to form the substrate for the majority of contemporary works.

At the end of the 20th century, two fields developed at the cutting edge of the world of one-dimensional photonic crystals: the field of inferential mirrors at the nanometric scale for X-ray and XUV, which filled a gap in the optical systems in this spectral field with many applications in spectroscopy and imaging, and the field of fiber Bragg gratings which marked a considerable advance in the field of telecommunications. This book is at one of those frontiers, in other words the application to the X-ray domain.

The creation of reliable artificial Bragg structures for the short wavelength domain began at the end of the 1970s. In fact, the use of such structures was envisaged for X-rays from the 1930 onwards by von Deubner (1930) and then by DuMond and Youtz (1935). The structures created (using silver) were unstable (due to interdiffusion), and it was not possible to use them in practical applications. Following the attempts of Dinklage (1967) and then Croce and Pardo (1970), it was not until the mid-1970s, when technologies allowing films to be deposited at the nanometric scale were mastered, that a few groups created effective Bragg mirrors in the XUV domain that remain relatively stable over time despite interdiffusion issues.

Spiller (1972), benefitting from the resources of the company IBM, can be cited as a pioneer in the development of these structures. Several research and academic institutions in various countries (mainly the United States, Germany, France, USSR) then joined the race. Therefore, we can mention the works by the Germans, in particular at DESY (synchrotron radiation center in Hamburg) (Haelbich and Kunz 1976), the works of a group of French laboratories led by Dhez and lastly the works of various Soviet laboratories (Gaponov et al. 1983).

Following an initial period of research and development, several private companies manufactured and marketed interferential X-ray mirrors, mainly in the United States (Ovonyx, etc.) and in Europe (LEP, Xenox, Incoatec, etc.), whereas institutional laboratories continued to manufacture structures generally for research purposes, sometimes outsourced and often at synchrotron radiation centers. Among these, we can note Troy Barbee's group in the United States, a group at the Institute of Theoretical and Applied Optics in France directed by Jean-Pierre Chauvineau and then Franck Delmotte, who worked with NASA for the solar telescope project SOHO, and the "Optics" group at the ESRF European synchrotron with Éric Ziegler and Andreas Freund.

At the same time, characterization devices were developed at synchrotron radiation centers; these were reflectometers operating in the field of soft X-ray: MOGOTOX at Super-ACO (Barchewitz and Marmoret 1988) followed by the metrology center installed at the SOLEIL synchrotron (Idir et al. 2006) in France, the Reflectometry Station at the Helmholtz Zentrum Berlin (Sokolov et al. 2014) and the metrology devices at the PTB laboratory installed at the BESSY II synchrotron in Germany, the BEAR station at the Elettra synchrotron in Italy (Nannarone et al. 2004), the CXRO reflectometer installed at the ALS synchrotron in the United States, etc. Other reflectometers installed in academic laboratories have been developed with sources of different kinds such as MONOX (André et al. 2005) or CEMOX (Hecquet et al. 2006).

Aside from these reflectometers, reflectometry operating with grazing incidence and generally using commercial equipment, has been widely used to characterize the type and quality of the interfaces between layers. Other techniques such as high-resolution X-spectrometry and mass spectrometry (Galtayries et al. 2010) or electronic microscopy (Häussler et al. 2007) have also been developed and have largely contributed to the understanding of the physical and chemical processes involved in creating these Bragg structures. Another important step forward in the field of artificial Bragg structures was obtained when different groups began to combine micro-manufacturing techniques used in microelectronics with manufacturing techniques used for Bragg X-ray mirrors.

From around the mid-1980s onwards, a group with Vitaly Aristov and Alexei Erko installed at Chernogolovka near Moscow (USSR) began to manufacture lenses known as Bragg–Fresnel lenses (Aristov et al. 1988), while in France at the LCP, Barchewitz and J.-M. André initiated manufacturing of lamellar multilayer gratings in Bragg mirrors (Barchewitz and André 1996). The purpose of the Bragg–Fresnel lenses was to focus the X-ray, whereas lamellar multilayer gratings allowed "high" spectral resolution Bragg monochromators and polychromators to be created. We will develop this subject in Chapter 7.

Since then, technologies have advanced unabated, enabling the deposition of nanometric thin layers on flat or curved optics, of high quality in terms of surfacing and polishing, to obtain a broad array of devices: mirrors, monochromators or lenses, intended for synchrotron light lines, for astrophysics, for X-ray microscopy, etc. The use of sliced multilayer structures at nanometric dimensions operating in transmission mode (Laue mode) has led to the creation of focusing lenses, with spatial resolution of the order of a few nanometers.

Aside from applications in optics, microscopy and spectroscopy, specific physical phenomena generated within Bragg structures have been discovered: the Purcell effect, Kossel diffraction, Bragg–Raman diffusion or topological effects. In addition, periodic multilayer structures have also been proposed and tested as elements, allowing X-rays to be generated according to certain principles: resonant transition radiation, parametric radiation, free electron laser, retro-distributed laser, etc. These various subjects will be discussed in Chapter 7.

1.3. References

Amiranoff, F. (2016). Apollon : le laser de l'extrême. *Reflets de la Physique*, 60–65.

André, J.-M., Maquet, A., Barchewitz, R. (1982). X-ray specular reflectivity and anomalous scattering in the vicinity of the Si K absorption edge in quartz. *Physical Review B*, 25(9), 5671–5679.

André, J.-M., Avila, A., Barchewitz, R., Benbalagh, R., Delaunay, R., Druart, D., Jonnard, P., Ringuenet, H. (2005). MONOX: A characterization tool for the X-UV range. *The European Physical Journal Applied Physics*, 31(2), 147–152.

Aristov, V.V., Erko, A.I., Martynov, V.V. (1988). Principles of Bragg-Fresnel multilayer optics. *Revue de Physique Appliquée*, 23(10), 1623–1630.

Authier, A. (2013). *Early Days of X-ray Crystallography*. Oxford University Press, Oxford.

Baldwin, G.C. and Kerst, D.W. (1975). Origin of synchrotron radiation. *Physics Today*, 28(1), 9–11.

Barchewitz, R. and André, J.-M. (1996). Structures multicouches gravées pour le rayonnement X-UV. *Le Journal de Physique IV*, 06(C4), C4-763–C4-772.

Barchewitz, R. and Marmoret, R. (1988). X-UV multilayer reflectivity tests using windowless soft X-rays tube and synchrotron source. *Revue de Physique Appliquée*, 23(10), 1661–1674.

Barchewitz, R., Montel, M., Bonnelle, C. (1967). Sélecteur de fréquences X continues réalisé pour le rayonnement d'orbite du synchrotron de Frascati. *Comptes Rendus Académie des Sciences Paris*, 264, 363–366.

Barchewitz, R., Bonnelle, C., Cremonese, M., Onori, G. (1969). Distribution spectrale du rayonnement X réfléchi par des miroirs dans un région d'absorption propre. *Comptes Rendus Académie des Sciences Paris*, 268, 151–154.

Barchewitz, R., Cremonese-Visicato, M., Onori, G. (1978). X-ray photoabsorption of solids by specular reflection. *Journal of Physics C: Solid State Physics*, 11(21), 4439.

Bessière, M. (1996). Introduction au rayonnement synchrotron et à ses avantages. *Journal de Physique IV Proceedings*, 06(C4), C4-511–C4-535.

Beye, M., Schreck, S., Sorgenfrei, F., Trabant, C., Pontius, N., Schüßler-Langeheine, C., Wurth, W., Föhlisch, A. (2013). Stimulated X-ray emission for materials science. *Nature*, 501(7466), 191–194.

Bloch, F. (1929). Über die Quantenmechanik der Elektronen in Kristallgittern. *Zeitschrift für Physik*, 52(7), 555–600.

Bonifacio, R., Pellegrini, C., Narducci, L.M. (1984). Collective instabilities and high-gain regime in a free electron laser. *Optics Communications*, 50(6), 373–378.

Bonnelle, C. and Dhez, P. (2014). Frascati : le rôle du Laboratoire de Chimie Physique dans les débuts de l'utilisation du rayonnement synchrotron. *Histoire de la recherche contemporaine. La revue du Comité pour l'histoire du CNRS*, III(1), 15–22.

Brillouin, L. (1946). *Wave Propagation In Periodic Structures*. McGraw-Hill, London [Online]. Available at: http://archive.org/details/dli.ernet.6377 [Accessed 25 November 2022].

Bruni, C., Jacquet, M., Loulergue, A., Monard, H., Zomer, F. (2016). ThomX : un démonstrateur de source compacte de rayons X. *Reflets de la Physique*, 50, 14–19.

Catoire, F., Quintard, L., Hort, O., Dubrouil, A., Constant, E. (2014). Les sources attosecondes. *Photoniques*, 28–33.

Cauchois, Y., Bonnelle, C., Missoni, G. (1963). Premiers spectres X du rayonnement d'orbite du synchrotron de Frascati. *Comptes Rendus Académie des Sciences Paris*, 257, 409–412.

Chapman, H.N., Barty, A., Bogan, M.J., Boutet, S., Frank, M., Hau-Riege, S.P., Marchesini, S., Woods, B.W., Bajt, S., Benner, W.H. et al. (2006). Femtosecond diffractive imaging with a soft-X-ray free-electron laser. *Nature Physics*, 2(12), 839–843.

Croce, P. and Pardo, B. (1970). Sur l'application des couches interférentielles à l'optique des rayons X et des neutrons. *Nouvelle Revue d'Optique Appliquée*, 1(4), 229–232.

Cukier, M., Dhez, P., Wuilleumier, F., Jaeglé, P. (1974). Photoionization cross section of uranium in the soft X-ray region. *Physics Letters A*, 48(4), 307–308.

von Deubner, W. (1930). Versuche über die Reflexion von Röntgenstrahlen an einem künstlich hergestellten Schichtenkörper. *Annalen der Physik*, 397(2), 261–280.

Dinklage, J.B. (1967). X-ray diffraction by multilayered thin-film structures and their diffusion. *Journal of Applied Physics*, 38(9), 3781–3785.

Drescher, M., Hentschel, M., Kienberger, R., Tempea, G., Spielmann, C., Reider, G.A., Corkum, P.B., Krausz, F. (2001). X-ray pulses approaching the attosecond frontier. *Science*, 291(5510), 1923–1927.

DuMond, J.W.M. and Youtz, J.P. (1935). Selective X-ray diffraction from artificially stratified metal films deposited by evaporation. *Physical Review*, 48(8), 703–703.

Elias, L.R., Fairbank, W.M., Madey, J.M.J., Schwettman, H.A., Smith, T.I. (1976). Observation of stimulated emission of radiation by relativistic electrons in a spatially periodic transverse magnetic field. *Physical Review Letters*, 36(13), 717–720.

Feldhaus, J., Saldin, E.L., Schneider, J.R., Schneidmiller, E.A., Yurkov, M.V. (1997). Possible application of X-ray optical elements for reducing the spectral bandwidth of an X-ray SASE FEL. *Optics Communications*, 140(4), 341–352.

Floquet, G. (1883). Sur les équations différentielles linéaires à coefficients périodiques. *Annales Scientifiques de l'École Normale Supérieure*, 12, 47–88.

Galtayries, A., Hu, M.-H., Le Guen, K., André, J.-M., Jonnard, P., Meltchakov, E., Hecquet, C., Delmotte, F. (2010). Nanometer-designed Al/SiC periodic multilayers: Characterization by a multi-technique approach. *Surface and Interface Analysis*, 42(6/7), 653–657.

Gaponov, S.V., Garin, F.V., Gusev, S.L., Kochemasov, A.V., Platonov, YY., Salashchenko, N.N., Gluskin, E.S. (1983). Multilayer mirrors for soft X-ray and VUV radiation. *Nuclear Instruments and Methods in Physics Research*, 208(1), 227–231.

Geloni, G., Kocharyan, V., Saldin, E. (2012). Self-seeding scheme for the soft X-ray line at the European XFEL [Online]. Available at: http://arxiv.org/abs/1202.5209 [Accessed 30 November 2022].

Gorkhover, T., Ulmer, A., Ferguson, K., Bucher, M., Maia, F.R.N.C., Bielecki, J., Ekeberg, T., Hantke, M.F., Daurer, B.J., Nettelblad, C. et al. (2018). Femtosecond X-ray Fourier holography imaging of free-flying nanoparticles. *Nature Photonics*, 12(3), 150–153.

Haelbich, R.-P. and Kunz, C. (1976). Multilayer interference mirrors for the XUV range around 100 eV photon energy. *Optics Communications*, 17(3), 287–292.

Häussler, D., Spiecker, E., Jäger, W., Störmer, M., Bormann, R., Michaelsen, C., Wiesmann, J., Zwicker, G., Benbalagh, R., André, J.-M. et al. (2007). Quantitative TEM characterizations of La/B4C and Mo/B4C ultrathin multilayer gratings by the geometric phase method. *Microelectronic Engineering*, 84(3), 454–459.

Hecquet, C., Roulliay, M., Delmotte, F., Ravet-Krill, M.F., Hardouin, A., Idir, M. (2006). Réflectomètre à large spectre EUV pour la métrologie d'optiques. *Journal de Physique IV (Proceedings)*, 138(1), 259–264.

Hoffmann, L., Jamnuch, S., Schwartz, C.P., Helk, T., Raj, S.L., Mizuno, H., Mincigrucci, R., Foglia, L., Principi, E., Saykally, R.J. et al. (2022). Saturable absorption of free-electron laser radiation by graphite near the carbon K-edge. *The Journal of Physical Chemistry Letters*, 13(39), 8963–8970.

Huang, Z. and Kim, K.-J. (2007). Review of X-ray free-electron laser theory. *Physical Review Special Topics - Accelerators and Beams*, 10(3), 034801.

Huang, N., Deng, H., Liu, B., Wang, D., Zhao, Z. (2021). Features and futures of X-ray free-electron lasers. *The Innovation*, 2(2), 100097.

Hyun, J., Satoh, M., Yoshida, M., Sakai, T., Hayakawa, Y., Tanaka, T., Hayakawa, K., Sato, I., Endo, K. (2018). Compact and intense parametric X-ray radiation source based on a linear accelerator with cryogenic accelerating and decelerating copper structures. *Physical Review Accelerators and Beams*, 21(1), 014701.

Idir, M., Brochet, S., Delmotte, A., Lagarde, B., Mercere, P., Moreno, T., Polack, F., Thomasset, M. (2006). Le pôle de métrologie de SOLEIL. *Journal de Physique IV (Proceedings)*, 138(1), 265–274.

Iwanenko, D. and Pomeranchuk, I. (1944). On the maximal energy attainable in a betatron. *Physical Review*, 65(11/12), 343–343.

Jaeglé, P. and Missoni, G. (1966). Coefficient d'absorption massique de l'or dans la région de longueur d'onde de 26 à 120 Å. *Comptes Rendus Académie des Sciences*, 262, 71–74.

Jaeglé, P., Missoni, G., Dhez, P. (1967). Study of the absorption of ultrasoft X rays by bismuth and lead using the orbit radiation of the Frascati synchrotron. *Physical Review Letters*, 19(1), 57.

Jonnard, P., André, J.-M., Le Guen, K., Wu, M., Principi, E., Simoncig, A., Gessini, A., Mincigrucci, R., Masciovecchio, C., Peyrusse, O. (2017). EUV stimulated emission from MgO pumped by FEL pulses. *Structural Dynamics*, 4(5), 054306.

Liénard, A. (1898). Champ électrique et magnétique produit par une charge électrique concentrée en un point et animée d'un mouvement quelconque. *L'éclairage électrique*, XVI(27), 5–14.

Madey, J.M.J. (1971). Stimulated emission of Bremsstrahlung in a periodic magnetic field. *Journal of Applied Physics*, 42(5), 1906–1913.

Matthews, D.L., Hagelstein, P.L., Rosen, M.D., Eckart, M.J., Ceglio, N.M., Hazi, A.U., Medecki, H., MacGowan, B.J., Trebes, J.E., Whitten, B.L. et al. (1985). Demonstration of a soft X-ray amplifier. *Physical Review Letters*, 54(2), 110–113.

Mottana, A. and Marcelli, A. (2013). Fifty years since the first European synchrotron radiation-derived XAFS spectrum (Frascati, 1963). *Journal of Synchrotron Radiation*, 20(5), 811–815.

Nannarone, S., Borgatti, F., DeLuisa, A., Doyle, B.P., Gazzadi, G.C., Giglia, A., Finetti, P., Mahne, N., Pasquali, L., Pedio, M. et al. (2004). The BEAR Beamline at Elettra. *AIP Conference Proceedings*, 705, 450–453.

Ostrowiecki, H. (2014). Le clin d'œil d'un expérimentateur. *Histoire de la recherche contemporaine. La revue du Comité pour l'histoire du CNRS*, III(1), 23–27.

Perrin, J. (1948). *La science et l'espérance.* Presses Universitaires de France, Paris [Online]. Available at: https://www.livre-rare-book.com/book/5472433/32366 [Accessed 25 November 2022].

Perrin, J. (1956). *Œuvres scientifiques*. CNRS Editions, Paris [Online]. Available at: https://www.cnrseditions.fr/catalogue/physique-et-astrophysique/oeuvres-scientifiques-jean-perrin/ [Accessed 28 November 2022].

Petroff, Y. and Thiry, P. (1980). Angle resolved photoemission in solids. *Applied Optics*, 19(23), 3957–3963.

Phillips, R.M. (1960). The Ubitron, a high-power traveling-wave tube based on a periodic beam interaction in unloaded waveguide. *IRE Transactions on Electron Devices*, 7(4), 231–241.

Piot, P., Brau, C.A., Gabella, W.E., Choi, B.K., Jarvis, J.D., Lewellen, J.W., Mendenhall, M.H., Mihalcea, D. (2012). R&D toward a compact high-brilliance X-ray source based on channeling radiation. *AIP Conference Proceedings*, 1507(1), 734–739.

Pollock, H.C. (1983). The discovery of synchrotron radiation. *American Journal of Physics*, 51(3), 278–280.

Prat, E. (2021). Synchrotron light sources and X-ray free-electron-lasers [Online]. Available at: http://arxiv.org/abs/2107.09131 [Accessed 30 November 2022].

Rayleigh, O.M. (1887). XVII. On the maintenance of vibrations by forces of double frequency, and on the propagation of waves through a medium endowed with a periodic structure. *The London, Edinburgh, and Dublin Philosophical Magazine and Journal of Science*, 24(147), 145–159.

Rayleigh, O.M. (1888). XXVI. On the remarkable phenomenon of crystalline reflexion described by Prof. Stokes. *The London, Edinburgh, and Dublin Philosophical Magazine and Journal of Science*, 26(160), 256–265.

Rohringer, N., Ryan, D., London, R.A., Purvis, M., Albert, F., Dunn, J., Bozek, J.D., Bostedt, C., Graf, A., Hill, R. et al. (2012). Atomic inner-shell X-ray laser at 1.46 nanometres pumped by an X-ray free-electron laser. *Nature*, 481(7382), 488–491.

Rulhusen, P., Artru, X., Dhez, P. (1998). *Novel Radiation Sources Using Relativistic Electrons.* World Scientific, Singapore [Online]. Available at: https://www.worldscientific.com/worldscibooks/10.1142/3398 [Accessed 9 January 2020].

Schwinger, J. (1949). On the classical radiation of accelerated electrons. *Physical Review*, 75(12), 1912–1925.

Sokolov, A.A., Eggenstein, F., Erko, A., Follath, R., Künstner, S., Mast, M., Schmidt, J.S., Senf, F., Siewert, F., Zeschke, T. et al. (2014). An XUV optics beamline at BESSY II. *Proceedings of SPIE*, 9206, 138–150.

Spiller, E. (1972). Low-loss reflection coatings using absorbing materials. *Applied Physics Letters*, 20(9), 365–367.

Stöhr, J. and Scherz, A. (2015). Creation of X-ray transparency of matter by stimulated elastic forward scattering. *Physical Review Letters*, 115(10), 107402.

Suckewer, S., Skinner, C.H., Milchberg, H., Keane, C., Voorhees, D. (1985). Amplification of stimulated soft X-ray emission in a confined plasma column. *Physical Review Letters*, 55(17), 1753–1756.

Thiry, P., Chandesris, D., Lecante, J., Guillot, C., Pinchaux, R., Pétroff, Y. (1979). E vs k and inverse lifetime of Cu(110). *Physical Review Letters*, 43(1), 82–85.

Togashi, T., Takahashi, E.J., Midorikawa, K., Aoyama, M., Yamakawa, K., Sato, T., Iwasaki, A., Owada, S., Okino, T., Yamanouchi, K. et al. (2011). Extreme ultraviolet free electron laser seeded with high-order harmonic of Ti:sapphire laser. *Optics Express*, 19(1), 317–324.

Tomboulian, D.H. and Hartman, P.L. (1956). Spectral and angular distribution of ultraviolet radiation from the 300-MeV Cornell synchrotron. *Physical Review*, 102(6), 1423–1447.

Weninger, C., Purvis, M., Ryan, D., London, R.A., Bozek, J.D., Bostedt, C., Graf, A., Brown, G., Rocca, J.J., Rohringer, N. (2013). Stimulated electronic X-ray Raman scattering. *Physical Review Letters*, 111(23), 233902.

Yablonovitch, E. and Gmitter, T.J. (1989). Photonic band structure: The face-centered-cubic case. *Physical Review Letters*, 63(18), 1950–1953.

Yoneda, H., Inubushi, Y., Nagamine, K., Michine, Y., Ohashi, H., Yumoto, H., Yamauchi, K., Mimura, H., Kitamura, H., Katayama, T. et al. (2015). Atomic inner-shell laser at 1.5-ångström wavelength pumped by an X-ray free-electron laser. *Nature*, 524(7566), 446–449.

Young, L., Kanter, E.P., Krässig, B., Li, Y., March, A.M., Pratt, S.T., Santra, R., Southworth, S.H., Rohringer, N., DiMauro, L.F. et al. (2010). Femtosecond electronic response of atoms to ultra-intense X-rays. *Nature*, 466(7302), 56–61.

2

Propagation of X-ray Radiation in a Bragg Structure: Frequency Domain

Jean-Michel ANDRÉ and Philippe JONNARD
LCPMR, CNRS, Sorbonne Université, Paris, France

2.1. Introduction to the interaction of X-rays and matter

X-ray radiation shows strong interaction with matter, primarily via the atomic electron cloud because the energy of photons, within the selected spectral domain, is of the order of the binding energy of the atomic electrons. The physics of these interactions is described mainly through quantum mechanics. The intensity of the sources of X-ray radiation currently available is such that the effects of the interaction are linear, even though at the time of writing, it is possible to observe nonlinear effects when free electron laser (XFEL) sources are used, in the X-ray domain. Effects of this type will not be discussed in this book.

From the point of view of optics, the X-ray–matter interaction is described by a quantity known as the optical constant, denoted as $\varepsilon(\boldsymbol{r})$, where $\boldsymbol{r}$ is the spatial variable and is independent of the time variable t. This constant will be considered a scalar quantity except when the optical medium is anisotropic in nature. It is assumed that the material medium reacts instantaneously with the electromagnetic field (no memory effect), which avoids the need for a time convolution when writing the displacement vector based on the dielectric constant and the electric field. It is

X-ray Radiation and Artificial Bragg Structures,
coordinated by Jean-Michel ANDRÉ and Philippe JONNARD.

also assumed that the nonlocality effects are negligible. The constant ε is a complex quantity ($\varepsilon \in \mathbb{C}$), and the introduction of the imaginary part is a practical way of rendering the attenuation effects. These are generated up to approximately 1 MeV, primarily by the photoelectric effect (photo-absorption) for the softer part of the spectrum and by Compton scattering for the hardest part.

The photoelectric effect corresponds to atomic excitation or ionization: under the impact of an X-ray photon, an electron from a given shell (K, L, M, etc.) may be ejected from this shell, provided that the energy of the photon is higher than the electron's binding energy. An absorption spectrum results, presenting discontinuities or absorption thresholds; Figure 2.1(a) shows a typical attenuation spectrum.

At the upper limit of the threshold (in other words, at photon energies slightly higher than the energy of the discontinuity), the absorption spectrum generally presents structures known as Kronig structures (de Laer Kronig 1931) and named X-ray absorption near edge structures (XANES) which extend several tens of eV beyond the discontinuity (see insets of Figure 2.1(a)), and oscillations known as extended X-ray absorption fine structures (EXAFS) which occur up to several hundred eV beyond the threshold (Lytle 1999). The first structures correspond to modulations in the density of the unoccupied electronic states that may receive the electron ejected during photo-absorption, in application of the selection rules of quantum mechanics. The second corresponds to scattering of the wave associated with the photoelectron, by the atoms adjacent to the atom from which the electron was ejected.

Compton scattering, which corresponds to an inelastic scattering of an X-ray photon by an electron, exercises an influence in terms of cross-section which depends on the atomic number and energy of the photon (see Figure 2.1(b)). Pair production only occurs in the very hard X-ray range and gamma radiation spectrum, spectral domains which are not concerned by Bragg optics.

At a basic level, the constant ε is obtained as the square of the optical index which we will write as $n^* = n \pm i\,\beta$[1]:

$$\varepsilon = (n \pm i\,\beta)^2 = n^2 - \beta^2 \pm i\,n\,\beta \qquad [2.1]$$

1. The sign depends on the selected orientation of the axes of the spatial frame of reference.

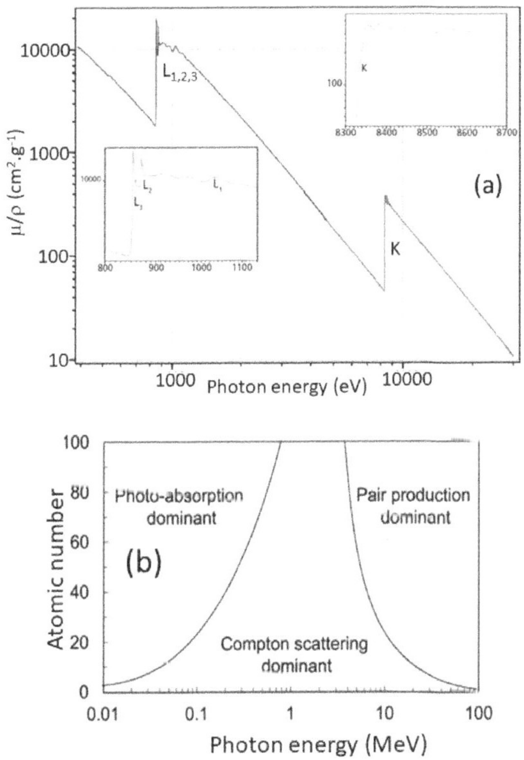

Figure 2.1. *a) Typical attenuation spectrum of X-rays; the case of nickel is illustrated here; figure adapted from Ménesguen et al. (2018); the insets show the detail of zones close to the K and L thresholds; b) relative importance of the different radiation–matter interactions; figure adapted from Parajuli et al. (2022)*[2]

The real part n describes the speed of propagation v of the wave (phase velocity) in the material medium and is defined by the ratio:

$$n = \frac{c}{v} \quad [2.2]$$

where c is the speed of light in a vacuum. Considering now the imaginary part β, this is linked to the linear absorption coefficient μ by the relationship:

$$\beta = \frac{\mu\,\lambda}{4\,\pi} \quad [2.3]$$

obtained from the optical theorem.

2. See doi: 10.3390/s22197374.

The real and imaginary parts n and β are not independent quantities. In fact, by virtue of the causality principle which stipulates that an effect cannot precede a cause, these two quantities are related by the so-called Kramers–Kronig relations (de Laer Kronig 1926; Kramers 1927) expressed below in terms of frequency as follows:

$$n(\omega) = 1 + \frac{c}{\pi} \int_0^{\infty} \frac{\beta(\Omega)}{\omega^2 - \Omega^2} d\Omega$$

$$\beta(\omega) = -\frac{4\omega^2}{\pi c} \int_0^{\infty} \frac{n(\Omega) - 1}{\omega^2 - \Omega^2} d\Omega \qquad [2.4]$$

It becomes apparent through these relations that the discontinuities of the absorption spectrum have an effect on dispersion of the real part; they cause significant variation of n around a threshold, constituting a phenomenon known as anomalous scattering. The absorption discontinuities and anomalous scattering of the index have considerable influence on the optical properties of the materials in near-threshold regions and consequently on the optical properties (reflectance, transmittance, bandwidth, etc.) of Bragg structures (Rosenbluth and Lee 1982; Marmoret and André 1983). Fine structures (XANES and EXAFS) also affect the spectrum of the real part n and can manifest themselves on the reflectance and transmittance spectra of simple mirrors (meaning bulk and with a homogeneous structure) (Barchewitz et al. 1978; André et al. 1982) and multilayer reflecting structures (Feng et al. 2021).

Very generally speaking, the real part of the optical index is slightly less than 1 in the X-ray domain, and we therefore write:

$$n = 1 - \delta \qquad [2.5]$$

The decrement values δ are of the order of 10^{-6} in the hard X-ray domain and of the order of 10^{-4} in the soft X-ray domain. The fact that $n < 1$ results in a total reflection phenomenon (or rather quasi-total reflection due to the absorption) effect for a grazing or glancing angle (further to the usual angle of incidence in optics) less than the critical angle which is approximately $\sqrt{2\delta}$. This value is obtained from the Snell–Descartes law of refraction. A consequence of this phenomenon is that Bragg mirrors are not spectrally selective for glancing angles. However, we can note that for certain light chemical elements near their K or L absorption threshold, the value of the real part n can be greater than 1 within a narrow spectral domain. This phenomenon gives rise to distinctive characteristics on the reflection spectrum (André and Barchewitz 1980) and can also result in a Cherenkov effect (Knulst et al. 2001; Kaplan and Shkolnikov 2005).

The real and imaginary parts of n^*, n and β respectively, which are macroscopic, are related to a microscopic quantity known as the atomic scattering factor, denoted as $f^* = f_1 + i\, f_2$ by the relation expressed in terms of frequency:

$$n^*(\omega) = 1 - 2\pi \left(\frac{c}{\omega}\right)^2 \sum_i N_i\, f_i^*.(\omega) \qquad [2.6]$$

The sum is made on the different types, i, of scattering centers present, of which there are N_i per unit volume.

Usually, the anomalous part in the expression for f_1 is separated out and it is written as $f_1 = f_0 + f'$; f_0 is identified in practical terms using the atomic number Z of the scattering atom. The anomalous term f' and the imaginary part f_2 can be calculated using the ab initio methods of relativistic quantum mechanics (Cromer and Liberman 1970) or semi-empirical methods (Parratt and Hempstead 1954; Kefi et al. 1992). The semi-empirical methods were widely used until data tables became available on the Internet, in particular those available on the website of the Center for X-Ray Optics (Berkeley, United States)[3].

2.2. Propagation in the frequency domain

We consider the theories that describe the propagation of X-ray radiation in artificial Bragg structures in the frequency domain from the point of view of those that apply to physical optics concerning one-dimensional photonic crystals. These Bragg structures are in practice made up of periodic superposition of a basic cell known as a unit cell with thickness d (see Figure 2.2). This unit cell is generally a bilayer structure but trilayer or quadrilayer cell structures have also been made. We note that given that the interface between two layers is generally not abrupt and perfectly smooth, an intermediate or transition layer between two main layers is commonly considered such that a nominally bilayer cell unit is in practice modeled as a trilayer structure.

The structure represented in Figure 2.2 can be of different forms. For example, the thickness can vary horizontally (lateral gradient of period) or vary according to the direction of stratification (depth gradient of thickness). In the following, more elaborate structures will be presented, made up of this type of basic structure. For example, we can mention lamellar gratings, blazed gratings or so-called Bragg–Fresnel lenses.

3. CXRO, X-Ray Interactions with Matter, see: http://henke.lbl.gov/optical_constants/.

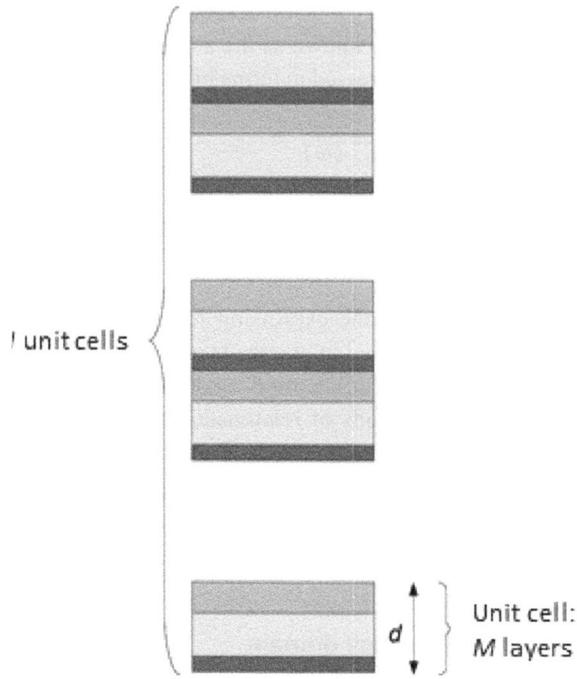

Figure 2.2. *Artificial Bragg structure made up of N unit cells with several layers; the unit cell is a trilayer (blue, yellow, red). For a color version of this figure, see www.iste.co.uk/andre/xrayradiation.zip*

Generally, the optics of photonic crystals "simply" uses propagation equations that govern spatial ($\boldsymbol{r}$) and temporal (t) changes in the electric field $\boldsymbol{E}(\boldsymbol{r},t)$ and magnetic field $\boldsymbol{H}(\boldsymbol{r},t)$, in a medium presenting a one-dimensional periodic optical modulation. In the absence of free charge and current, these are second-order differential equations (not coupled), deduced from the Maxwell equations, which in a linear optical, isotropic, non-magnetic material medium, characterized by a dielectric scalar constant ε, are written in the Gauss unit system as:

$$\boldsymbol{\nabla}^2\boldsymbol{E}(\boldsymbol{r},t)-\frac{\varepsilon(\boldsymbol{r})}{c^2}\frac{\partial^2\boldsymbol{E}(\boldsymbol{r},t)}{\partial t^2}+\boldsymbol{\nabla}\left(\boldsymbol{E}(\boldsymbol{r},t).\left(\frac{\boldsymbol{\nabla}\,\varepsilon(\boldsymbol{r})}{\varepsilon(\boldsymbol{r})}\right)\right)=\boldsymbol{0} \quad [2.7]$$

$$\boldsymbol{\nabla}^2\boldsymbol{H}(\boldsymbol{r},t)-\frac{\varepsilon(\boldsymbol{r})}{c^2}\frac{\partial^2\boldsymbol{H}(\boldsymbol{r},t)}{\partial t^2}+\left(\frac{\boldsymbol{\nabla}\,\varepsilon(\boldsymbol{r})}{\varepsilon(\boldsymbol{r})}\right)\wedge\left(\boldsymbol{\nabla}\wedge\boldsymbol{H}(\boldsymbol{r},t)\right)=\boldsymbol{0} \quad [2.8]$$

2.2.1. *Master equations and the Hill equation*

While the fields are described by monochromatic plane waves, the derivation with respect to time in equations [2.6] and [2.7] results in the following equation for a wave with a transverse electric field:

$$\boldsymbol{\nabla} \wedge \boldsymbol{\nabla} \wedge \boldsymbol{E_k} - \varepsilon\, k^2\, \boldsymbol{E_k} = \boldsymbol{0}$$

so:

$$\boldsymbol{\nabla}^2 \boldsymbol{E_k} + \varepsilon\, k^2\, \boldsymbol{E_k} = \boldsymbol{0} \qquad [2.9]$$

and the following for a wave with a transverse magnetic field:

$$\boldsymbol{\nabla} \wedge \boldsymbol{\nabla} \wedge \boldsymbol{H_k} - \frac{\nabla\varepsilon}{\varepsilon} \wedge \boldsymbol{\nabla} \wedge \boldsymbol{H_k} - \varepsilon\, k^2\, \boldsymbol{H_k} = \boldsymbol{0}$$

so:

$$\boldsymbol{\nabla} \wedge \left[\frac{1}{\varepsilon}\, \boldsymbol{\nabla} \wedge \boldsymbol{H_k}\right] + k^2\, \boldsymbol{H_k} = \boldsymbol{0} \qquad [2.10]$$

These equations are sometimes called Helmholtz equations. They do not contain time dependence and therefore apply to the frequency domain. In our situation, the dielectric constant only varies according to a single spatial dimension ($\varepsilon(\boldsymbol{r}) = \varepsilon(z)$, where z is perpendicular to the surface of the layers). As long as the medium is isotropic, in order for $\varepsilon(z)$ to remain a scalar quantity, the electric field and the magnetic field are divided into a so-called transverse electric wave, *TE,* and a so-called transverse magnetic field, *TM,* by making use of the tangential component conservation relation $k_{\|}$ (perpendicular to the z axis) of the wave vector. This conservation corresponds to the Snell–Descartes law, explicitly:

$$k_{\|}{}^2 + k_z(z)^2 = \varepsilon(z)\, k^2 \qquad [2.11]$$

For *TE* polarization, we write:

$$E^{TE}{}_x(z) = U(z)\, exp(i\, k_{\|}\, y)\,, E^{TE}{}_y(z) = 0, E^{TE}{}_z(z) = 0$$

$$H^{TE}{}_x(z) = 0, H^{TE}{}_y(z) = V(z)\, exp(i\, k_{\|}\, y)\,, H^{TE}{}_z(z) = W(z)\, exp(i\, k_{\|}\, y),$$

$$[2.12]$$

and for *TM* polarization:

$$H^{TM}{}_x(z) = U(z)\,exp(i\,k_{\parallel}\,y)\,, H^{TM}{}_y(z) = 0, H^{TM}{}_z(z) = 0$$

$$E^{TM}{}_x(z) = 0, E^{TM}{}_y(z) = -V(z)\,exp(i\,k_{\parallel}\,y)\,, E^{TM}{}_z(z) = \mathrm{W}(z)\,exp(i\,k_{\parallel}\,y)$$

[2.13]

To simplify, the index $\boldsymbol{k}$ has been omitted from the expression for amplitude of the fields. The amplitudes $U(z)$, $V(z)$ and $W(z)$ are connected by the Maxwell equations. By introducing the previous expressions of these fields in the propagation equations, the following propagation equations are obtained:

$$\frac{d^2U(z)}{dz^2} + h(z)^2U(z) = 0, \qquad [2.14]$$

where for *TE* polarization:

$$h(z) = k_z(z) \qquad [2.15]$$

and for *TM* polarization:

$$h(z)^2 = k_z(z)^2 - \frac{1}{\varepsilon(z)}\frac{d\,\varepsilon(z)}{dz}\frac{d}{dz}. \qquad [2.16]$$

If we write $F(z) = U(z)$ and $h(z) = k_z(z)$ for the *TE* case, and $F(z) = U(z)/n(z)$ and $h(z)^2 = k_z(z)^2 + \frac{1}{\varepsilon(z)^2}\frac{d^2n(z)}{dz^2} + \frac{1}{\varepsilon(z)^3}\frac{d\,\varepsilon(z)}{dz}\frac{d\,n(z)}{dz}$ for the *TM* case, then we obtain the following unified propagation equation for the two polarizations:

$$\frac{d^2F(z)}{dz^2} + h(z)^2F(z) = 0 \qquad [2.17]$$

Taking into account the periodic nature of the optical index, it follows that the quantity $h(z)^2$ is also periodic (with the same period as $\varepsilon(z)$) and that consequently equation [2.17] is a Hill equation (1886). It is fundamental for studying the propagation of an electromagnetic wave in a one-dimensional artificial Bragg structure, a 1D-PC. Various approaches can be used to resolve this equation. If the dimension of the 1D-PC is infinite, therefore in practice very large compared to the wavelength, the Bloch–Floquet method implemented in dynamic diffraction of X-ray radiation by a crystal or in physics of crystalline solids to study the electronic structure can be used: the simplest, so-called two-wave, version is effective in the case of strong coupling; otherwise, a multi-wave approach must be used.

2.2.2. *Solving the Hill equation using the Bloch–Floquet method*

The fundamental basis for this method is a theorem by Floquet (1883), which was updated by Bloch in solid state physics. As seen above, propagation of an electromagnetic wave within a 1D-PC in the frequency domain is governed by a Hill equation, with a solution of the form:

$$F(z) = f(z)\, exp(i\, K\, z) \qquad [2.18]$$

The function $f(z)$ is a periodic function with period d, and K is a quantity to be determined according to the problem, and which in the context of optics is generally known as the Bloch–Floquet wave constant or quasi-moment. The periodic function $f(z)$ will be expanded into a Fourier series:

$$f(z) = \sum_{n=-\infty}^{+\infty} f_n\, exp(i\, n\, g\, z); n \in \mathbb{Z} \qquad [2.19]$$

where $g = \frac{2\,\pi}{d}$ is the reciprocal wave number. The Fourier coefficients f_n are spatial harmonics of the Bloch wave which have no intrinsic and independent existence but are just components of the overall wave. Similarly, the term $h^2(z)$ for equation [2.17] has the following Fourier expansion:

$$h^2(z) = k^2 \sum_{p=-\infty}^{+\infty} J_p\, exp(i\, p\, g\, z);\ p \in \mathbb{Z} \qquad [2.20]$$

When these Fourier expansions for $f(z)$ and $h^2(z)$ are introduced into the Hill equation, an infinite system of coupled algebraic equations is obtained, which can be summarized as follows:

$$f_n \ = \ k^2 \sum_{m=-\infty}^{+\infty} f_m \frac{J_{n-m}}{(K+n\, g)^2 - k^2\, J_0};\ m \in \mathbb{Z}^*\ (m \neq 0) \qquad [2.21]$$

This system of equations has a non-trivial solution provided that the Cramer determinant (of infinite dimension), associated with this system and denoted as $\Delta(K)$, is zero. The problem was studied by G.W. Hill in the context of his study of movement of the Moon (Hill 1886). Applying his approach, it results that the Bloch–Floquet number must fulfill the relation:

$$sin^2\left(K\ \frac{d}{2}\right) = \ \Delta(0)\ sin^2\left(k\ \frac{\sqrt{J_0}\ d}{2}\right) \qquad [2.22]$$

with:

$$\Delta(0) = \Delta(K) \text{ and } K = 0 \qquad [2.23]$$

The conditions for convergence of this type of infinite determinant were studied by H. Poincaré. Numerical simulations show that convergence is rapid in practical cases, and that it is generally not necessary to go further than the dimension 15×15 in order to achieve three-figure precision. An important case in practice is a sinusoidal profile for which, in the Fourier expansion, only the terms $p = 0, \pm 1$ are retained; it corresponds to the so-called Mathieu equation, which is a specific case of the Hill equation.

Equation [2.22] is in fact a dispersion relation which can be considered an equation with the Bloch–Floquet number K as its unknown. Real value solutions (in terms of K) lead to stable states, whereas complex values give unstable states. Physically, stable solutions correspond to progressive modulated waves, whereas unstable solutions are damped or growing waves. The latter, which do not satisfy the radiation condition at infinity, are therefore physically excluded. Lastly, we note that this relation appears to be a generalization of the second Snell–Descartes law to a 1D-PC, with $\sqrt{\Delta(0)}$ taking on the role of the relative optical index and the quantities below the sinus taking on the roles of the angle of incidence θ_i and angle of refraction θ_r.

The quantity J_0 that appears in [2.22] corresponds to an apparent dielectric constant:

$$J_0 = \bar{\epsilon} - \sin^2 \theta_i \qquad [2.24]$$

given that the quantity $\bar{\epsilon}$ is the average dielectric constant. Detailed analysis of equation [2.22] clarifies the situation and allows the various modes of propagation (André and Pardo 1988) to be determined. This analysis is inspired by the second Snell–Descartes law: the case where K is imaginary gives rise to damped, or evanescent, waves, a phenomenon corresponding to extinction in the dynamic theory of diffraction. Table 2.1 summarizes the results in the case of media where no loss or gain occurs.

The different cases are identified using matrix-type notation (1.1, 1.2, etc.). The first column (cases 1.1 and 2.1) corresponds to positive apparent dielectric constants J_0, and to Bragg selective reflection: the system of inequalities represented in square 2.1 constitutes a rigorous expression of the Bragg reflection conditions. The second column corresponds to negative values of J_0. Case 2.2 corresponds to total reflection: the structure is equivalent to a homogeneous medium with a dielectric constant equal to $\bar{\epsilon}$ and in a total reflection situation. Case 2.1 also corresponds to a total reflection situation; however, the additional condition indicated for reflection to occur must be met; otherwise, the structure will present transmission windows. These transmission windows correspond to guided modes inside the structure.

	$J_0 > 0$	$J_0 < 0$	X
$\Delta(0) < 0$	$Re(K) = 2\,p\,\frac{g}{2}$ no additional condition 1.1	$Re(K) = (2\,p + 1)\,\frac{g}{2}$ $sinh^2(X)\,\lvert\Delta(0)\rvert > 1$ 1.2	$X = k\,\frac{d}{2}\sqrt{J_0}$
$\Delta(0) > 0$	$Re(K) = (2\,p + 1)\frac{g}{2}$ $sin^2(X)\,\Delta(0) > 1$ 2.1	$Re(K) = 2\,p\,\frac{g}{2}$ no additional condition 2.2	$X = k\,\frac{d}{2}\sqrt{J_0}$

Table 2.1. *Propagation regimes according to the law of dispersion [2.22]*

Cases 1.1 and 2.1 correspond to the stopbands (also referred to as bandgaps or simple gaps) of even (case 1.1) or odd (case 2.1) orders. These phenomena are usually illustrated by a dispersion diagram in the frequency ω/Bloch–Floquet wave number K plane, such as the one represented in Figure 2.3(a). The frequency ω_0 is the Bragg frequency, given by:

$$\omega_O = \frac{g}{2}\frac{c}{\sqrt{J_0}} \quad [2.25]$$

The frequencies ω_{0u} and ω_{0d} respectively denote the frequencies of the upper and lower edges of the stopband, the width of which is denoted by the term Δ_B. The dispersion curves $\omega(K)$ are located within so-called Brillouin zones, where the number K is between $n\ g/2$ and $(n+1)\ g/2$, where n is a positive integer or zero. We should recall that the number K is a complex number; in a non-absorbing and passive medium, it presents an imaginary part only in the stopband (Bragg zone between ω_{0u} and ω_{0l}), which results in the phenomenon of extinction. For absorbing or active media, the number K has an imaginary non-zero value outside the stopbands (true absorption or gain).

At the limit of the Brillouin zone, at the edge of the bands, the derivatives of the dispersion curves $\frac{d\omega}{dK}$ tend to cancel each other out, which corresponds to the phenomenon of sub-luminal propagation. In the stopband, a system of stationary waves establishes itself in the structure. This situation is shown in Figure 2.3(b) for the case of a bilayer stratified medium with band-edge frequencies u and d. The spatial position of the wave changes when the frequency $\omega(K)$ moves through the stopband. In the middle of the band, the nodes are located almost at the interfaces. This "sliding" of the stationary wave is used in certain analysis techniques for determining the physical and chemical states of the structure.

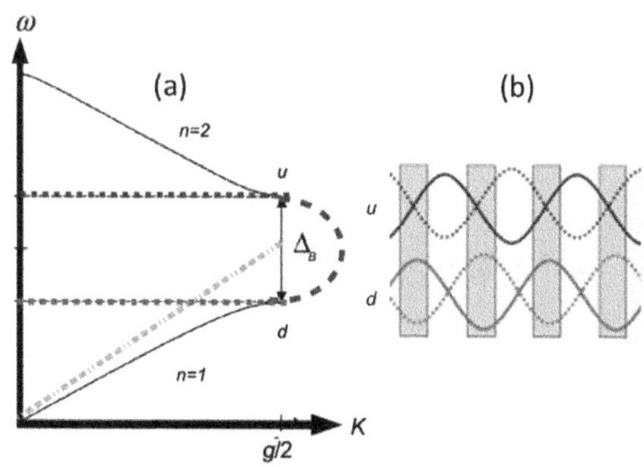

Figure 2.3. *Dispersion diagram. For a color version of this figure, see www.iste.co.uk/andre/xrayradiation.zip*

COMMENT ON FIGURE 2.3.– *a) Dispersion diagram showing only the first and second bands (n = 1 and 2). The thin black solid curves correspond to the real part of the Bloch–Floquet number (quasi-moment) K, whereas the dotted purple curve corresponds to the imaginary part in the case of a non-absorbing and passive medium. The green dotted line represents the light line (no dispersion).* Δ_B *is the width of the stopband; b) stationary waves in a stratified bilayer medium in correspondence with the band edges: (u) blue curves, (d) red curves.*

Once the wave number K has been determined, the system of equations [2.21] can be resolved to give the relative values of the harmonics f_n / f_0, where f_0 has been obtained from the limit conditions. In the particular case of the Mathieu equation, corresponding to a sinusoidal profile, the system of equations is reduced to:

$$D_n f_n + f_{n+1} + f_{n-1} = 0;\ D_n = 2\,\frac{f_0 - (K+n\,g)^2}{f_1} \qquad [2.26]$$

Applying an iterative method to relation [2.21], the amplitude ratios of the harmonics f_n / f_{n+1} are expressed using continuous fractions:

$$D_n = \frac{1}{D_{n-1} - \frac{1}{D_{n-2} - \frac{1}{D_{n-3}}\cdots}} + \frac{1}{D_{n+1} - \frac{1}{D_{n+2} - \frac{1}{D_{n+3}}\cdots}} \qquad [2.27]$$

In many cases, coupling between the fundamental ($n = 0$) and a particular order is dominant, let us say $n = p$. In this case, the problem will be a two-wave problem and we have:

$$f_p = k^2 f_0\,\frac{J_p}{(K+p\,g)^2 - k^2 J_0} \qquad [2.28]$$

To illustrate this idea, let us suppose that coupling is dominant between the fundamental 0 and the order +1. The determinant of the coupled system of dimension 2 must cancel itself out to obtain non-trivial solutions, which leads to the equation:

$$[K^2 - k^2 J_0][(K + g)^2 - k^2 J_0] = k^4\,(J_1 J_{-1})^2 \qquad [2.29]$$

of which the approximate solutions are given by:

$$K = \frac{g}{2} \pm \sqrt{\left(k\sqrt{J_0} + \frac{g}{2}\right)^2 - \frac{k^4\,(J_1 J_{-1})^2}{g^2}} \qquad [2.30]$$

The Hill–Floquet approach is instructive from the point of view of physics of electromagnetic wave propagation in one-dimensional periodic media: it predicts the existence of stopbands corresponding to the Bragg diffraction domains and allows their width to be estimated. Strictly speaking, this approach is only valid for an infinite medium. In practice, semi-infinite media or media of finite dimensions should be examined, and in this case, to obtain exact numerical estimates, other methods must be used. Two main methods can be identified: matrix methods and recursive methods.

2.2.3. *Matrix methods*

Matrix methods are based on the linear nature of the propagation problem. Historically, the first use of matrix form in thin-film optics is generally attributed to Herpin (1947). The matrices involved will of course depend on the (vectorial) space in which they will act and on the selected polarization. In practice, a different space will be used: the space of horizontal components (in other words, tangential to the plane of stratification) of the electric and magnetic fields, which are continuous quantities at the interfaces, which is the point of view adopted in the well-known works of Abelès (1950a, 1950b); the space formed from a component of a field and its derivative with respect to the space variable z, which is the approach considered by Lekner (1994); or the space of the amplitudes of direct or retrograde waves, which is frequently used in the literature, in particular in the formalization of transfer matrices, and allows a more direct physical discussion than that possible with the so-called Abelès method. Other approaches that avoid numerical stability problems have been developed, such as scattering matrix (S matrix) approaches or R matrix approaches.

2.2.3.1. *Abelès method and Lekner approach*

In the first space listed above, in other words the space of the horizontal components of the fields, the amplitudes denoted as $(U(z), V(z))$ (see equations [2.12] and [2.13] for the notations) verify the following relation in accordance with the Maxwell equations:

$$\begin{pmatrix} U'(z) \\ V'(z) \end{pmatrix} = i\,k_0 \begin{pmatrix} 0 & \zeta \\ \frac{1}{\zeta}\frac{k_z^2}{k_0^2} & 0 \end{pmatrix} \begin{pmatrix} U(z) \\ V(z) \end{pmatrix} \equiv \widehat{D} \begin{pmatrix} U(z) \\ V(z) \end{pmatrix} \qquad [2.31]$$

where $\zeta = 1$ for *TE* polarization and $\zeta = \varepsilon$ for *TM* polarization. The symbol "$'$" denotes here the derivation with respect to the variable z.

In a layer comprising a homogeneous medium, the matrix $\widehat{D}$ is independent of z. In a layer of this kind, the column vector $\begin{pmatrix} U(z) \\ V(z) \end{pmatrix}$ at position z is connected to the column vector $\begin{pmatrix} U(z_0) \\ V(z_0) \end{pmatrix}$ at position z_0 in the same layer by the propagator $\hat{g}$ that validates:

$$\hat{g}(z, z_0) = \hat{g}(z)\hat{g}^{-1}(z_0) \qquad [2.32]$$

so:

$$\begin{pmatrix} U(z) \\ V(z) \end{pmatrix} = \hat{g}(z, z_0) \begin{pmatrix} U(z_0) \\ V(z_0) \end{pmatrix} \qquad [2.33]$$

We obtain:

$$\frac{d}{dz}\,\hat{g}(z, z_0) = \widehat{D}\,\hat{g}(z, z_0) \qquad [2.34]$$

Integration of this differential equation gives:

$$\hat{g}(z) = exp\left(\hat{D}\, z\right) \qquad [2.35]$$

A simple algebraic calculation leads to:

$$\hat{g}(z) = \begin{pmatrix} cos(k_z z) & \frac{i}{\xi}\, sin(k_z z) \\ i\, \xi\, sin(k_z z) & cos(k_z z) \end{pmatrix} \qquad [2.36]$$

with $\xi = \frac{{k_z}^2}{\zeta^2\, k_0^2}$.

The matrix $\hat{g}(z)$ is symplectic, with a determinant equal to 1. Using the operator $\hat{g}(z)$, we can construct the characteristic matrix for the layer j of thickness d_j:

$$\widehat{M_j} = \begin{pmatrix} cos\left(k_{z_j} d_j\right) & \frac{i}{\xi_j}\, sin\left(k_{z_j} d_j\right) \\ i\, \xi_j\, sin\left(k_{z_j} d_j\right) & cos\left(k_{z_j} d_j\right) \end{pmatrix} \qquad [2.37]$$

At this stage, we point out that the characteristic group of matrices $\widehat{M_j}$ equipped with the matrix multiplication " . " is not Abelian; in fact, the product $\widehat{M_h}.\widehat{M_k}$ is not commutative. We have:

$$\widehat{M_h}.\widehat{M_k} - \widehat{M_k}.\widehat{M_h} = \left[\widehat{M_h}, \widehat{M_k}\right] = \hat{X}_{hk} \qquad [2.38]$$

with

$$\hat{X}_{hk} = \begin{pmatrix} \left(\frac{\xi_h}{\xi_k} - \frac{\xi_k}{\xi_h}\right) sin\left(k_{z_h} d_h\right) cos\left(k_{z_k} d_k\right) & 0 \\ 0 & \left(\frac{\xi_k}{\xi_h} - \frac{\xi_h}{\xi_k}\right) sin\left(k_{z_h} d_h\right) cos\left(k_{z_k} d_k\right) \end{pmatrix}$$

$$[2.39]$$

This type of relation can be expanded to:

$$\widehat{M_h}.\widehat{M_k}.\widehat{M_l} - \widehat{M_l}.\widehat{M_k}.\widehat{M_h} = \frac{1}{2}\left\{\left\{\widehat{M_h}, \left[\widehat{M_k}, \widehat{M_l}\right]\right\} - \left\{\widehat{M_k}, \left[\widehat{M_l}, \widehat{M_h}\right]\right\} + \left\{\widehat{M_l}, \left[\widehat{M_h}, \widehat{M_k}\right]\right\}\right\} \qquad [2.40]$$

where:

$$\left\{\widehat{M_h}, \left[\widehat{M_k}, \widehat{M_l}\right]\right\} = \widehat{M_h}.\left[\widehat{M_k}, \widehat{M_l}\right] + \left[\widehat{M_k}, \widehat{M_l}\right].\widehat{M_h} \qquad [2.41]$$

where $\left[\widehat{M_k},\widehat{M_l}\right]$ is the anticommutator. We note that $\hat{X}_{hk}$, and therefore the anticommutators, are diagonal matrices with zero trace, properties that are involved in the question of reciprocity.

A 1D-PC is formed from repetition of a unit cell associated with the characteristic matrix $\hat{C}_{unit}$. For a stratified medium of which the unit cell is formed of M layers, the characteristic matrix $\hat{C}_{unit}$ can be obtained as the product of the characteristic matrices of the M layers forming it. An overall characteristic matrix $\widehat{C}_g$ is associated with the 1D-PC, this characteristic matrix having been obtained as the product of the N characteristic matrices of the N unit cells:

$$\widehat{C}_g = \left(\hat{C}_{unit}\right)^N \equiv \begin{pmatrix} c_{g11} & c_{g12} \\ c_{g21} & c_{g22} \end{pmatrix} \quad [2.42]$$

We will see that given the symplectic nature of the matrix $\hat{C}_{unit}$, the product will, by virtue of the Cayley–Hamilton theorem, be expressed compactly using Chebychev polynomials for periodic systems (crystals but not for quasi-crystals). The matrix $\widehat{C}_g$ allows the tangential components to be connected at the substrate, *sub,* to those at the level of the incident medium, *inc,* and therefore to calculate the reflectance $R = |r_N|^2$ and the transmittance $T = \frac{k_{zsub}}{k_{zinc}}|t_N|^2$ using the relationships:

$$t_N = \frac{2\,\xi_{inc}}{\xi_{inc}\,c_{g11}+\xi_{sub}\,c_{g22}+\xi_{inc}\,\xi_{sub}\,c_{g12}+c_{g21}}$$

$$r_N = \frac{\xi_{inc}\,c_{g11}-\xi_{sub}\,c_{g22}+\xi_{inc}\,\xi_{sub}\,c_{g12}-c_{g21}}{\xi_{inc}\,c_{g11}+\xi_{sub}\,c_{g22}+\xi_{inc}\,\xi_{sub}\,c_{g12}+c_{g21}} \quad [2.43]$$

The case of non-absorbing or passive media (in other words, with no loss or gain) is interesting to examine. For this type of case, the transmittance is written as:

$$T = \frac{4\,\xi_{inc}\,\xi_{sub}}{\left(\xi_{inc}\,c_{g11}+\xi_{sub}\,c_{g22}\right)^2-\left(\xi_{inc}\,\xi_{sub}\,c_{g12}+c_{g21}\right)^2} \quad [2.44]$$

with the reflectance given, in accordance with the energy conservation law, by:

$$R = 1 - T \quad [2.45]$$

We can note that the transmittance can be rewritten in the form:

$$T \equiv \frac{1}{\Upsilon^2-\mho^2} \quad [2.46]$$

where in the particular case where $\xi_{inc} = \xi_{subs} = 1$, Υ is associated with the matrix trace $\widehat{C}_g$ and $\mho$ with its anti-trace.

The method described above is F. Abelès's well-known method (Abelès 1950a, 1950b). The characteristic matrices involved in this method are, for non-absorbing and passive dielectric media, complex with real diagonal elements and imaginary extra-diagonal elements. However, the elements of the matrices that connect a field and its derivative with respect to the spatial variable *z* (e.g. $E^{TE}{}_x(z)$ and $\frac{dE^{TE}{}_x(z)}{dx}$ in the case of *TE* polarization) are real in the same case, which can be an advantage from a numerical point of view. This is the point of view of Lekner (1994).

Let us consider the *TE* case. According to the theory of second-order differential equations, the solution to the propagation equation related to $E^{TE}{}_x(z)$ is written in the form:

$$E^{TE}{}_x(z) = f\,F(z) + g\,G(z) \qquad [2.47]$$

where *f* and *g* are two constants. The component $E^{TE}{}_x$ and its derivative $\frac{dE^{TE}{}_x}{dx}$ at a point z_j^- are connected to the component $E^{TE}{}_x$ and its derivative $\frac{dE^{TE}{}_x}{dx}$ at a point z_j^+ by a matrix with dimension 2 denoted as $\widehat{\mathcal{M}_j}$:

$$\begin{pmatrix} E^{TE}{}_x \\ \frac{dE^{TE}{}_x}{dx} \end{pmatrix}_{z_j^-} = \widehat{\mathcal{M}_j} \begin{pmatrix} E^{TE}{}_x \\ \frac{dE^{TE}{}_x}{dx} \end{pmatrix}_{z_j^+} \qquad [2.48]$$

In addition, the component $E^{TE}{}_x$ and its derivative $\frac{dE^{TE}{}_x}{dx}$ at the point z_j^+ are connected by the matrix $\widehat{\mathrm{A}_j}$ to the constants *f* and *g*:

$$\begin{pmatrix} E^{TE}{}_x \\ \frac{dE^{TE}{}_x}{dx} \end{pmatrix}_{z_j^+} = \widehat{\mathrm{A}_j} \begin{pmatrix} f \\ g \end{pmatrix} \qquad [2.49]$$

with:

$$\widehat{\mathrm{A}_j} = \begin{pmatrix} F(z_j^+) & G(z_j^+) \\ F'(z_j^+) & G'(z_j^+) \end{pmatrix} \qquad [2.50]$$

with the prime " ′ " used here to denote the derivation with respect to *z*. Repeating the operation at point z_j^- with a matrix $\widehat{\mathrm{B}_j}$:

$$\begin{pmatrix} E^{TE}{}_x \\ \frac{dE^{TE}{}_x}{dx} \end{pmatrix}_{z_j^-} = \widehat{\mathrm{B}_j} \begin{pmatrix} f \\ g \end{pmatrix} \qquad [2.51]$$

Introducing [2.49], [2.50] and [2.51] into [2.48] gives:

$$\widehat{\mathcal{M}_J} = \widehat{\mathrm{B}_J}\, \widehat{\mathrm{A}_J}^{-1} \qquad [2.52]$$

The matrix $\widehat{\mathcal{M}_J}$ is symplectic, given that its determinant is equal to 1.

Stratified structures are an important specific case. For these, the dielectric constant has a constant value in a given layer. In this case, we can write:

$$\widehat{\mathcal{M}_J} = \begin{pmatrix} cos(\delta_j) & \frac{1}{k_{zj}} sin(\delta_j) \\ - k_{zj}\, sin(\delta_j) & cos(\delta_j) \end{pmatrix} \qquad [2.53]$$

The quantity δ_j corresponds to the optical phase shift as the layer j is traversed.

In the same way as for the method of characteristic matrices, taking into account the symplectic nature of the matrix $\widehat{\mathcal{M}_J}$ and in accordance with the Cayley–Hamilton theorem, the global matrix of the structure $\widehat{\mathcal{M}}_g$ may be obtained using Chebychev polynomials for periodic systems. The reflectance $R = |r_N|^2$ and transmittance $T = \frac{k_{zsub}}{k_{zinc}} |t_N|^2$ can be expressed by means of the elements in the overall matrix.

2.2.3.2. *Transfer matrices*

Inside a 1D-PC, the electric field or the magnetic field can be written as the superposition of a direct wave $\mathcal{F}(z) = \mathcal{F}\, exp(i\, k_z\, z)$ and a retrograde wave $\mathcal{R}(z) = \mathcal{R}\, exp(-\, i\, k_z z)$. These two waves form the spacc on which so-called transfer matrices will act. In the representation $\begin{pmatrix} \mathcal{F}(z) \\ \mathcal{R}(z) \end{pmatrix}$, the propagator $\widehat{G}(z)$ is obviously given by the diagonal matrix $\begin{pmatrix} exp(i\, k_z\, z) & 0 \\ 0 & exp(-\, i\, k_z z) \end{pmatrix}$. The similarity matrix $\widehat{S}$, which transforms $\begin{pmatrix} \mathcal{F}(z) \\ \mathcal{R}(z) \end{pmatrix}$ into $\begin{pmatrix} U(z) \\ V(z) \end{pmatrix}$, corresponding to a change of base, is obtained from the eigensystem for $\widehat{g}(z)$, that is, $\Lambda^{\pm} = \exp(\pm\, i\, k_z\, z)$ for eigenvalues and $\boldsymbol{e}^{\pm} = \begin{pmatrix} 1 \\ \pm\xi \end{pmatrix}$ for eigenvectors, which gives for $\hat{S}$:

$$\widehat{S} = (\boldsymbol{e}^{+}, \boldsymbol{e}^{-}) = \begin{pmatrix} 1 & 1 \\ +\xi & -\xi \end{pmatrix} \qquad [2.54]$$

Contrary to the fields $\begin{pmatrix} U(z) \\ V(z) \end{pmatrix}$, the fields $\begin{pmatrix} \mathcal{F}(z) \\ \mathcal{R}(z) \end{pmatrix}$ are not continuous at an interface and there is a transition matrix denoted as $\widehat{P}$, which connects $\begin{pmatrix} \mathcal{F}(z) \\ \mathcal{R}(z) \end{pmatrix}^{+}$ from one side of the interface in the layer $j+1$ to $\begin{pmatrix} \mathcal{F}(z) \\ \mathcal{R}(z) \end{pmatrix}^{-}$ on the other side of the interface in layer j:

$$\begin{pmatrix} \mathcal{F}(z) \\ \mathcal{R}(z) \end{pmatrix}^{+} = \widehat{P} \begin{pmatrix} \mathcal{F}(z) \\ \mathcal{R}(z) \end{pmatrix}^{-} \quad [2.55]$$

We obtain $\widehat{P}$ from the following path:

– transformation of $\begin{pmatrix} \mathcal{F}(z) \\ \mathcal{R}(z) \end{pmatrix}$ into $\begin{pmatrix} U(z) \\ V(z) \end{pmatrix}$ in the medium j by the matrix $\widehat{S}_j$;

– traversal of the interface in the representation $\begin{pmatrix} U(z) \\ V(z) \end{pmatrix}$ by the identity matrix as a result of the continuity of the tangential components;

– transformation of $\begin{pmatrix} U(z) \\ V(z) \end{pmatrix}$ into $\begin{pmatrix} \mathcal{F}(z) \\ \mathcal{R}(z) \end{pmatrix}$ in the medium $j+1$ by the matrix $\widehat{S}^{-1}$ which gives $\widehat{P} = \left(\widehat{S}_{j+1}\right)^{-1} \widehat{S}_j$, so:

$$\widehat{P}_{j,j+1} = \frac{1}{2} \begin{pmatrix} 1 + \frac{\xi_{j+1}}{\xi_j} & 1 - \frac{\xi_{j+1}}{\xi_j} \\ 1 - \frac{\xi_{j+1}}{\xi_j} & 1 + \frac{\xi_{j+1}}{\xi_j} \end{pmatrix} \quad [2.56]$$

The matrix $\widehat{P}_{j,j+1}$ is in fact constructed from the reflection $r_{j,j+1}$ and transmission $t_{j,j+1}$ coefficients of the fields at the interface and, to this end, can be considered a matrix that generalizes the Fresnel formulation, such that $\widehat{P}_{j,j+1}$ is sometimes called the Fresnel matrix.

The coefficients $r_{j,j+1}$ and $t_{j,j+1}$ are obtained by supposing that only one wave is transmitted in the medium $j+1$. We obtain:

$$r_{j,j+1} = \frac{\xi_j - \xi_{j+1}}{\xi_{j+1} + \xi_j}$$

$$t_{j,j+1} = \frac{2\,\xi_j}{\xi_{j+1} + \xi_j} \quad [2.57]$$

such that $\widehat{P}_{j+1,j}$ can be rewritten using these coefficients as follows:

$$\widehat{P}_{j,j+1} = \frac{1}{t_{j,j+1}} \begin{pmatrix} 1 & r_{j,j+1} \\ r_{j,j+1} & 1 \end{pmatrix} \quad [2.58]$$

For the unit cell formed from M layers, the corresponding propagation matrix $\widehat{\wp}_u$ is made up of the product of the matrices corresponding to each layer $\widehat{P}_{s,s+1}\,\widehat{G}_s$. The overall propagation matrix $\widehat{\wp}_g$ of the N unit cells is obtained by raising $\widehat{\wp}_u$ to the power N:

$$\widehat{\wp}_g = (\widehat{\wp}_u)^N = \begin{pmatrix} \wp_{g11} & \wp_{g12} \\ \wp_{g21} & \wp_{g22} \end{pmatrix} \quad [2.59]$$

The operation of raising to the power N can be performed using a method based on the Cayley–Hamilton theorem presented below. The elements of $\widehat{\wp}_g$ allow the reflection and transmission coefficients to be obtained in each extreme medium:

$$r^- = \left.\frac{\mathcal{R}^-}{\mathcal{F}^-}\right]_{\mathcal{R}^+=0} = -\frac{\wp_{g21}}{\wp_{g22}},\ t^- = \left.\frac{\mathcal{F}^+}{\mathcal{F}^-}\right]_{\mathcal{R}^+=0} = \frac{det[\widehat{\wp}_g]}{\wp_{22}}$$

$$r^+ = \left.\frac{\mathcal{F}^+}{\mathcal{R}^+}\right]_{\mathcal{F}^-=0} = +\frac{\wp_{g12}}{\wp_{g22}},\ t^+ = \left.\frac{\mathcal{R}^-}{\mathcal{R}^+}\right]_{\mathcal{F}^-=0} = \frac{1}{\wp_{g22}} \quad [2.60]$$

Since we have:

$$\begin{pmatrix} \mathcal{F}^+(0) \\ \mathcal{R}^+(0) \end{pmatrix} = \widehat{\wp}_g \begin{pmatrix} \mathcal{F}^-(L) \\ \mathcal{R}^-(L) \end{pmatrix} \quad [2.61]$$

the matrix $\widehat{\wp}_g$ can be written using these coefficients as follows:

$$\widehat{\wp}_g = \frac{1}{t^+} \begin{pmatrix} \Theta & r^+ \\ -r^- & 1 \end{pmatrix} \quad [2.62]$$

with:

$$\Theta \equiv t^+t^- - r^+r^- \quad [2.63]$$

Under certain conditions (related to symmetries, transparency and passivity) $\Theta = 1$. In this case, the transmission coefficient is independent of the incidence (right or left), whereas the reflection coefficient remains a priori dependent on the incidence.

2.2.3.3. *Reflectance and transmittance from the Bloch phase*

The Cayley–Hamilton theorem tells us that a matrix $\widehat{\mathcal{M}}$ satisfies its own characteristic equation, so for a symplectic matrix:

$$\widehat{\mathcal{M}}^2 = Tr[\widehat{\mathcal{M}}]\,\widehat{\mathcal{M}} - \mathbb{I}_2 \qquad [2.64]$$

Inducing from the previous equation, raising a symplectic matrix to the power *N* can be obtained in a compact form:

$$\widehat{\mathcal{M}}^N = \frac{1}{sin\phi}\left(\widehat{\mathcal{M}}\, sin(N\phi) - \mathbb{I}_2 sin((N-1)\phi)\right) \qquad [2.65]$$

where the angle ϕ has been introduced by $cos\phi = \frac{1}{2}\, Tr[\widehat{\mathcal{M}}]$. It follows that $\widehat{\mathcal{M}}^N$ is written using second-order Chebychev polynomials $U_N(\phi) = \frac{sin(\phi(N+1))}{sin(\phi)}$ as follows:

$$\begin{pmatrix} \mathcal{M}_{11} & \mathcal{M}_{12} \\ \mathcal{M}_{21} & \mathcal{M}_{22} \end{pmatrix}^N = \begin{pmatrix} \mathcal{M}_{11}\, U_{N-1}(\phi) - U_{N-2}(\phi) & \mathcal{M}_{12}\, U_{N-1}(\phi) \\ \mathcal{M}_{21}\, U_{N-1}(\phi) & \mathcal{M}_{22}\, U_{N-1}(\phi) - U_{N-2}(\phi) \end{pmatrix} \qquad [2.66]$$

This formula applies to both the representation $\begin{pmatrix} U(z) \\ V(z) \end{pmatrix}$ and the representation $\begin{pmatrix} \mathcal{F}(z) \\ \mathcal{R}(z) \end{pmatrix}$. From the latter, the reflection coefficient is obtained directly as a function of the Bloch phase ϕ:

$$r^-{}_N = -\frac{\wp_{g21}}{\wp_{g22}} = -\frac{\wp_{u21}\, U_{N-1}(\phi)}{\wp_{u22}\, U_{N-1}(\phi) - U_{N-2}(\phi)} \qquad [2.67]$$

from which follows the reflectance $R^- = |r^-{}_N|^2$.

Since:

$$\frac{r^-{}_N}{t^-{}_N} = \wp_{g21} \qquad [2.68]$$

$$\frac{r^-{}_1}{t^-{}_1} = \wp_{u21} \qquad [2.69]$$

several relationships can be deduced (Vezzetti and Cahay 1986):

$$(T^-)^{-1} \equiv (|t^-{}_N|^2)^{-1} = 1 + |U_{N-1}(\phi)|^2 \left(\frac{1}{|t^-{}_1|^2} - 1\right) \qquad [2.70]$$

$$R^- = |r^-{}_1|^2\, |1 + U_{N-2}(\phi)\, t^-{}_N|^2 \qquad [2.71]$$

The second-order Chebychev polynomials $U_N(\phi)$ are rapidly oscillating functions with *K* (Bloch–Floquet number) for non-absorbing media and, in this case, they "print" their behavior onto R^- and T^-. Therefore, in a quasi-transparent medium, between two stopbands, there are *N-1* zeros and *N-2* lateral reflectance lobes, with the reflectance peak located at the center of the stopband. Similar relations can be obtained mutatis mutandis for R^+ and T^+.

As an example of the transfer matrices method and of application of the method of raising a matrix to a power, presented previously, we show in Figure 2.4 the reflectance and the transmittance at normal incidence as a function of the radiation energy of an interferential stratified (self-supported) mirror composed of 150 bilayers formed from one layer of lanthanum *La* with thickness 1.41 nm and a layer of boron carbide B_4C with thickness 1.97 nm. The optical indices used are taken from the references Palik (1985) and André et al. (2005). We note that as a result of the existence of the "true" absorption phenomenon (as opposed to extinction), the peak reflectance does not reach the value 1 and the Darwin plateau tends to disappear. We also note that secondary peaks appear next to the main peak (Bragg peak), which correspond to interferences between "super-cells" formed by the association of 2, 3, ..., bilayers. Generally, there are therefore *N*-2 secondary peaks between each Bragg peak for a mirror which has *N* bilayers.

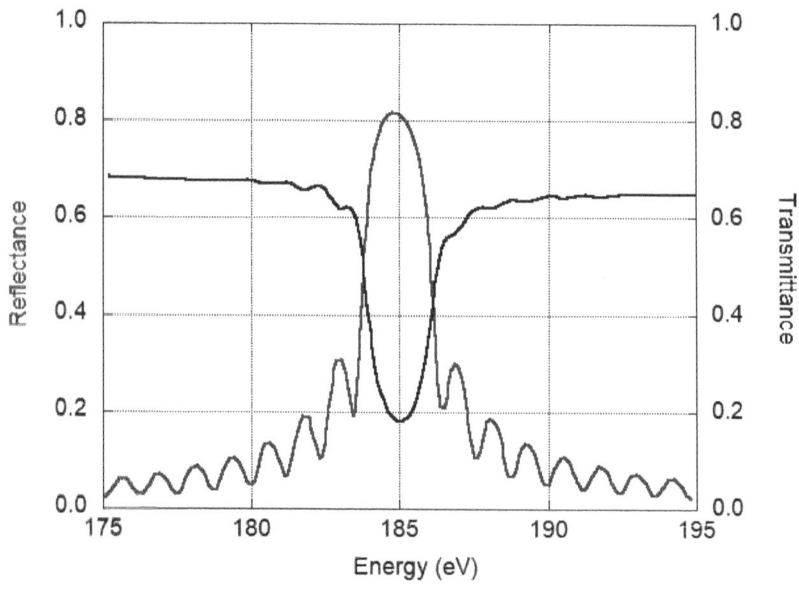

Figure 2.4. *Calculation using the matrix method for reflectance (in red) and transmittance (in blue) at normal incidence of an interferential mirror La/B$_4$C as a function of the radiation energy. For a color version of this figure, see www.iste.co.uk/andre/xrayradiation.zip*

2.2.3.4. *Dispersion relation and Bloch–Floquet number as a function of the elements in the transfer matrix*

According to the Floquet theorem, there is a number K (Bloch–Floquet number) which verifies:

$$\begin{pmatrix} \mathcal{F}(z) \\ \mathcal{R}(z) \end{pmatrix}_{j+1} = \exp(i\,K\,d) \begin{pmatrix} \mathcal{F}(z) \\ \mathcal{R}(z) \end{pmatrix}_{j} \quad [2.72]$$

Strictly speaking, this relation is exact for a 1D-PC of infinite dimensions. But for systems that are not very absorbent and are passive with a large number of unit cells, it remains a satisfactory approximation. Given this, we also have:

$$\begin{pmatrix} \mathcal{F}(z) \\ \mathcal{R}(z) \end{pmatrix}_{j+1} = \widehat{\wp}_u \begin{pmatrix} \mathcal{F}(z) \\ \mathcal{R}(z) \end{pmatrix}_{j} \quad [2.73]$$

which leads us to say that $\exp(i\,K\,d)$ is an eigenvalue of $\widehat{\wp}_u$ and is given by:

$$\exp(i\,K\,d) = \frac{1}{2}\,Tr(\widehat{\wp}_u) \pm \sqrt{\left\{\left[\frac{1}{2}\,Tr(\widehat{\wp}_u)\right]^2 - 1\right\}} \quad [2.74]$$

The dispersion relation follows:

$$K\,d = Arccos\left[\frac{1}{2}\,Tr(\widehat{\wp}_u)\right] \quad [2.75]$$

We write:

$$\cos(\phi_u) = \frac{1}{2}\,Tr(\widehat{\wp}_u) \quad [2.76]$$

If we restrict ourselves to media with a real refraction index, therefore non-absorbing and passive, we can distinguish three regimes:

– $\left|\frac{1}{2}\,Tr(\widehat{\wp}_u)\right| < 1$: K is real and the waves are progressive Bloch waves; the 2 eigenvalues of $\widehat{\wp}_u$: μ_1,μ_2 belong to the unit complex circle and are conjugated complex numbers.

– $\left|\frac{1}{2}\,Tr(\widehat{\wp}_u)\right| > 1$: K is imaginary and the waves are evanescent Bloch waves; the corresponding frequency domain ω is called the stopband. The 2 eigenvalues of $\widehat{\wp}_u$ are real, one with a modulus greater than 1; let us say $|\mu_2| > 1$ and therefore $|\mu_1| < 1$.

– $\left|\frac{1}{2}\, Tr(\widehat{\wp}_u)\right| = 1$: corresponds to the band threshold regime.

In the usual case of a stratified medium formed from bilayers made up of one layer 1 with thickness d_1 and a dielectric constant ε_1 and a layer 2 with thickness d_2 and a dielectric constant ε_2, the dispersion equation takes the form:

$$\cos(K\, d) = cos(k_{z1}\, d_1)\, cos(k_{z2}\, d_2) - \frac{1}{2}\left(\frac{\varepsilon_1\, k_{z2}}{\varepsilon_2\, k_{z1}} + \frac{\varepsilon_2\, k_{z1}}{\varepsilon_1\, k_{z2}}\right) sin(k_{z1}\, d_1)\, sin(k_{z2}\, d_2)$$

[2.77]

Still in the case of a bilayer system, a non-explicit calculation of the corresponding matrix $\widehat{\wp}_u$ shows that the trace of this matrix is given by:

$$Tr(\widehat{\wp}_u) = \frac{2}{1-\left|r_{2,1}\right|^2}\left(cos(\Phi_{2,1}) - \left|r_{2,1}\right|^2 sin\,(\Delta_{2,1})\right) \qquad [2.78]$$

with:

$$\Phi_{2,1} = k_{z1}\, d_1 + k_{z2}\, d_2, \Delta_{2,1} = k_{z1}\, d_1 - k_{z2}\, d_2 \qquad [2.79]$$

Using this last relation, we can show that the width of the stopband Δk in terms of the wave number is proportional to $\left|r_{2,1}\, sin(p\, \gamma\, \pi)\right|$, with p being the order of the stopband (order of Bragg diffraction). As a result, the following extinction rules are obtained:

– The stopband disappears if $r_{2,1}$ cancels out, in a similar way to the Brewster angle for *TM* polarization.

– The stopband disappears if $p\, \gamma\, \pi$ is a multiple of π; thus, there is no stopband with an even order if $\gamma = 1/2$, of order 3 if $\gamma = \frac{1}{3}$ or $\frac{2}{3}$, ..., in other words the condition for extinction of the orders is:

$$\gamma = \frac{k}{p}, k = 1, \ldots, p-1 \qquad [2.80]$$

In the most complicated case of a stratified medium formed from several layers ($M > 2$), whether in the case of a periodic (crystal) or aperiodic (quasi-crystal) structure, the product of the matrices turns out to be more difficult to calculate.

2.2.3.5. *S Matrices*

The construction of transfer matrices connects the amplitudes of the fields on either side of an element in the system (interface, layer, unit cell, etc.); thus, for matrix $\hat{P}$:

$$\begin{pmatrix}\mathcal{F}\\\mathcal{R}\end{pmatrix}^+ \equiv \begin{pmatrix}\mathcal{F}^+\\\mathcal{R}^+\end{pmatrix} = \hat{P}\begin{pmatrix}\mathcal{F}\\\mathcal{R}\end{pmatrix}^- \equiv \hat{P}\begin{pmatrix}\mathcal{F}^-\\\mathcal{R}^-\end{pmatrix} \qquad [2.81]$$

However, it is possible to connect the different amplitudes using a different "matrix organization". Therefore, we can inverse the order in the column matrices using T matrices:

$$\begin{pmatrix}\mathcal{R}^+\\\mathcal{F}^+\end{pmatrix} = \hat{T}\begin{pmatrix}\mathcal{R}^-\\\mathcal{F}^-\end{pmatrix} \qquad [2.82]$$

It is also possible to introduce scattering matrices (S matrices) which connect outgoing fields to incoming fields:

$$\begin{pmatrix}\mathcal{R}^+\\\mathcal{F}^-\end{pmatrix} = \hat{\Sigma}\begin{pmatrix}\mathcal{F}^+\\\mathcal{R}^-\end{pmatrix} \qquad [2.83]$$

We note that the form used for S matrices denoted as $\hat{\Sigma}$ can differ from one scientific community to another, with the field of optics (Li 1996) adopting a formalism that is often different from the conventional formalism of the community in the field of microwaves. This question and its implications have been discussed by Rumpf (2011). Although the conventional formalism offers certain advantages, we will present the form generally adopted by opticians for which the elements $\Sigma_{11}$and Σ_{22} correspond to transmission terms while the elements Σ_{12}, Σ_{21} correspond to reflection terms.

The elements in matrices $\hat{\Sigma}$ can be connected to those in the transfer matrices $\hat{P}$; thus, by a simple algebraic equation, we obtain:

$$\Sigma_{11} = P_{11} - \frac{P_{21}P_{12}}{P_{22}} = \frac{\det[\hat{P}]}{P_{22}};\ \Sigma_{12} = \frac{P_{12}}{P_{22}};\ \Sigma_{21} = -\frac{P_{21}}{P_{22}};\ \Sigma_{22} = \frac{1}{P_{22}} \qquad [2.84]$$

To treat a stratified system, it is necessary to combine the $\hat{\Sigma}$ matrices related to each layer to obtain the overall matrix. Contrary to the case of transfer matrices or T matrices, this operation is not performed using standard matrix multiplication, but instead using a specific product denoted as $\otimes$ and generally referred to as the Redheffer star product, defined as follows:

$$\hat{\Sigma}^{(AB)} \equiv \begin{pmatrix} \Sigma^{(AB)}{}_{11} & \Sigma^{(AB)}{}_{12} \\ \Sigma^{(AB)}{}_{21} & \Sigma^{(AB)}{}_{22} \end{pmatrix} = \hat{\Sigma}^{(A)} \otimes \hat{\Sigma}^{(B)} =$$
$$\begin{pmatrix} \Sigma^{(A)}{}_{11} & \Sigma^{(A)}{}_{12} \\ \Sigma^{(A)}{}_{21} & \Sigma^{(A)}{}_{22} \end{pmatrix} \otimes \begin{pmatrix} \Sigma^{(B)}{}_{11} & \Sigma^{(B)}{}_{12} \\ \Sigma^{(B)}{}_{21} & \Sigma^{(B)}{}_{22} \end{pmatrix} \quad [2.85]$$

with:

$$\Sigma^{(AB)}{}_{11} = \Sigma^{(A)}{}_{11} \Sigma^{(B)}{}_{11} \left(1 - \Sigma^{(B)}{}_{21} \Sigma^{(A)}{}_{12}\right)^{-1}$$

$$\Sigma^{(AB)}{}_{22} = \Sigma^{(A)}{}_{22} \Sigma^{(B)}{}_{22} \left(1 - \Sigma^{(B)}{}_{21} \Sigma^{(A)}{}_{12}\right)^{-1}$$

$$\Sigma^{(AB)}{}_{12} = \Sigma^{(B)}{}_{12} + \Sigma^{(B)}{}_{11} \Sigma^{(B)}{}_{22} \Sigma^{(A)}{}_{12} \left(1 - \Sigma^{(B)}{}_{21} \Sigma^{(A)}{}_{12}\right)^{-1}$$

$$\Sigma^{(AB)}{}_{21} = \Sigma^{(A)}{}_{21} + \Sigma^{(B)}{}_{21} \Sigma^{(A)}{}_{22} \Sigma^{(A)}{}_{11} \left(1 - \Sigma^{(B)}{}_{21} \Sigma^{(A)}{}_{12}\right)^{-1} \quad [2.86]$$

The Redheffer star product is associative, a property that can be put to good use to improve the calculation time.

The algorithm to determine the overall matrix $\hat{\Sigma}_g$ is as follows: $\hat{\Sigma}_g$ is firstly initialized by writing $\Sigma_{g11} = \Sigma_{g22} = 1$; $\Sigma_{g12} = \Sigma_{g21} = 0$; then, the matrix is calculated for the first layer $\hat{\Sigma}^{(1)}$ and the new overall matrix is generated by $\hat{\Sigma}_g = \hat{\Sigma}_g \otimes \hat{\Sigma}^{(1)}$. Finally, the operation is repeated for the next *N*–*1* layers such that:

$$\hat{\Sigma}_g = \hat{\Sigma}^{(N)} \otimes \hat{\Sigma}^{(N-1)} \otimes \ldots \hat{\Sigma}^{(1)} \quad [2.87]$$

For a periodic system, therefore for a crystal and not a quasi-crystal, we can use the "doubling" algorithm: we construct the S matrix for the unit cell $\hat{\Sigma}_u$, as we explained for the overall matrix, and then we convert the decimal number of unit cells of the crystal into a binary number. With each digit equal to 1 is associated the relevant S matrix corresponding to the power of 2, while for each digit equal to 0, the relevant S matrix corresponding to the power of 2 is eliminated from the calculation. To illustrate this, let us consider the following example: a crystal has 102 cell units; the corresponding binary number is 1100110, and the associated S matrix is given by:

$$\hat{\Sigma}_{102} = (1 \times \hat{\Sigma}_{64}) \otimes (1 \times \hat{\Sigma}_{32}) \otimes (0 \times \hat{\Sigma}_{16}) \otimes (0 \times \hat{\Sigma}_{8}) \otimes (1 \times \hat{\Sigma}_{4}) \otimes (1 \times \hat{\Sigma}_{2}) \otimes (0 \times \hat{\Sigma}_{1}) \quad [2.88]$$

so:

$$\hat{\Sigma}_{102} = \hat{\Sigma}_{64} \otimes \hat{\Sigma}_{32} \otimes \hat{\Sigma}_{4} \otimes \hat{\Sigma}_{2} \quad [2.89]$$

In the previous case:

$$\hat{\Sigma}_1 = \hat{\Sigma}_u, \hat{\Sigma}_2 = \hat{\Sigma}_1 \otimes \hat{\Sigma}_1, \hat{\Sigma}_4 = \hat{\Sigma}_2 \otimes \hat{\Sigma}_2, \ldots, \hat{\Sigma}_{64} = \hat{\Sigma}_{32} \otimes \hat{\Sigma}_{32} \quad [2.90]$$

Therefore, the algorithm using S matrices, associated with the doubling method, constitutes an effective way of dealing with the case of a crystal with a large number of unit cells, which are themselves made up of a high number of layers, a system that is often difficult to manage using transfer matrices.

S matrices were initially used in particle scattering theory and in the theory of electric circuits (quadripoles), in particular for transmission lines, which enabled this field to move closer to the field of optics in stratified media. These matrices became "popular" in the 1950s and 1960s in the field of electromagnetism with the appearance of microwave network analyzers introduced by the company Hewlett Packard. From a numerical point of view, the use of S matrices avoids the stability problems that transfer matrices or T matrices sometimes present. Indeed, as soon as the optical indices become complex, the elements of transfer matrices or T matrices can contain numbers that are both very small and very large, corresponding to exponentially evanescent or growing waves, all the more so given that the structure is thick. There is a resulting numerical instability. During numerical calculation by a computer of finite precision, small quantities can be set to zero (loss of significant digits) and large quantities can exceed calculation capacities. It is apparent that a reduction in the thickness of each layer without reducing the total thickness does not allow this problem to be resolved because the value of the exponential terms increases when the matrices are multiplied.

2.2.3.6. *R matrices*

Aside from the scattering matrices method, a method has been developed in which so-called R matrices act on the horizontal components (U, V). This approach is to the Abelès method of characteristic matrices, which act on (U, V), what the scattering matrices method is to the method of transfer matrices, which act on $(\mathcal{F}^+, \mathcal{R}^+)$. R matrices $\hat{R}$ connect the fields (U, V) in two adjacent layers *j, j–1* via the relation:

$$\begin{pmatrix} U(z) \\ U(z) \end{pmatrix}_{z_j} = \hat{R} \begin{pmatrix} V(z) \\ V(z) \end{pmatrix}_{z_{j-1}} \quad [2.91]$$

The elements of matrices $\hat{R}$ can be connected to those of transfer matrices $\hat{P}$; thus, using simple algebra, we obtain:

$$R_{11} = \frac{P_{11}}{P_{21}};\ R_{12} = P_{12} - \frac{P_{22}P_{11}}{P_{21}};\ R_{21} = \frac{1}{P_{21}};\ R_{22} = -\frac{P_{22}}{P_{21}} \qquad [2.92]$$

When dealing with a multilayer system, it is necessary to combine the matrices $\hat{R}$ relative to each layer to obtain the overall matrix. In the same way as scattering matrices, this operation is not performed using standard matrix multiplication, but instead using a specific R-type product that we will denote $\circledast$ (to be differentiated from the Redheffer star product defined in [2.85]) and defined as follows:

$$\hat{R}^{(AB)} \equiv \begin{pmatrix} R^{(AB)}{}_{11} & R^{(AB)}{}_{12} \\ R^{(AB)}{}_{21} & R^{(AB)}{}_{22} \end{pmatrix} = \hat{R}^{(A)} \circledast \hat{R}^{(B)} = \begin{pmatrix} R^{(A)}{}_{11} & R^{(A)}{}_{12} \\ R^{(A)}{}_{21} & R^{(A)}{}_{22} \end{pmatrix} \circledast \begin{pmatrix} R^{(B)}{}_{11} & R^{(B)}{}_{12} \\ R^{(B)}{}_{21} & R^{(B)}{}_{22} \end{pmatrix} \qquad [2.93]$$

with:

$$R^{(AB)}{}_{11} = R^{(B)}{}_{11} + \Sigma^{(B)}{}_{21}\Sigma^{(B)}{}_{12}\left(R^{(B)}{}_{22} - R^{(A)}{}_{11}\right)^{-1}$$

$$R^{(AB)}{}_{22} = R^{(A)}{}_{22} + \Sigma^{(A)}{}_{21}\Sigma^{(A)}{}_{12}\left(R^{(B)}{}_{22} - R^{(A)}{}_{11}\right)^{-1}$$

$$R^{(AB)}{}_{12} = \Sigma^{(B)}{}_{12}\Sigma^{(A)}{}_{12}\left(R^{(B)}{}_{22} - R^{(A)}{}_{11}\right)^{-1}$$

$$R^{(AB)}{}_{21} = \Sigma^{(A)}{}_{21}\Sigma^{(B)}{}_{21}\left(R^{(B)}{}_{22} - R^{(A)}{}_{11}\right)^{-1} \qquad [2.94]$$

The product $\circledast$ is also associative. The algorithm to determine the overall matrix $\hat{R}_g$ is as follows: $\hat{R}_g$ is firstly initialized by writing $R_{g11} = 1$; $R_{g22} = -1$; $R_{g12} = R_{g21} = 0$; then, we proceed in a similar way to the algorithm for scattering matrices. The method of R matrices is also numerically stable. We point out that other combinations are also possible: thus, use the idea of scattering matrices (connecting incoming waves to outgoing waves) but acting on fields (U, V) as seen in the R' matrices approach of Montiel and Nevière (1994).

2.2.3.7. *Problem of coherence*

In the previous sections, propagation of the radiation is assumed to be coherent. However, there are situations in which propagation takes on an incoherent or partially coherent nature; this is the case in particular for sources that do not emit in phase. Different approaches have been proposed to deal with this situation: the square of the fields, or corrective factors, is used on Fresnel coefficients (Katsidis and Siapkas 2002). It has also been proposed that a random phase should be used, which is added to the geometric phase φ_g, of graduated importance to pass from coherence to incoherence (Troparevsky et al. 2010); thus, the phase φ is written in the form:

$$\varphi = \varphi_g + \beta\, Rand \qquad [2.95]$$

where β takes on values between 0 and π and where $Rand$ is a random number between −1 and +1. Partial coherence is represented by:

$$0 < \beta < \pi \qquad [2.96]$$

and incoherence by $\beta = \pi$.

2.2.4. *Recursive methods*

In matrix methods, in particular the method of transfer matrices and T matrices, numerical instability problems can arise, as previously pointed out. Therefore, regions of total reflection are sensitive to this problem. In this case, it can be helpful to use recursive methods. There are several of them: one that is frequently used by those working with optics in the X-ray field is known as the Parratt method; others are methods based on impedance or admittance; and lastly there are those that use laws of composition such as generalization of Einstein's velocity addition law. This list is not exhaustive but presents some methods that we will describe in more detail. We will point out that these methods may apply to stratified media that are not necessarily periodic, which is useful when dealing with the case of structures that are not perfectly periodic, either accidentally (fabrication defect, interdiffusion, etc.) or artificially, to obtain specific structures (super-mirrors, mirrors with a period gradient, etc.), which resemble photonic quasi-crystals.

2.2.4.1. *Parratt method*

This calculation procedure is an extension of the method initially developed by L.G. Parratt to deal with the effects of surface contamination in reflectometry of X-ray radiation at a low glancing angle (Parratt 1954). It is well-adapted to dealing with the case of stratified stacks that are not necessarily periodic, such as that shown in diagram form in Figure 2.5.

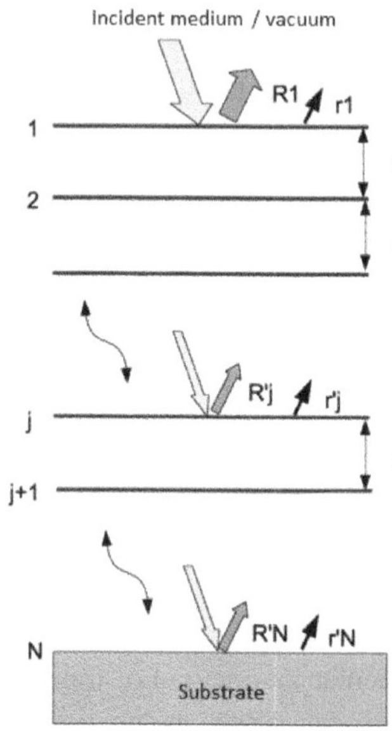

Figure 2.5. *Diagram of a stratified structure (not necessarily periodic) composed of N layers arranged on a substrate; each layer j has a thickness d_j; at each interface, the radiation is reflected by all the underlying layers with a coefficient R'; the Fresnel reflection coefficient at the interface j is denoted as r'j. For a color version of this figure, see www.iste.co.uk/andre/xrayradiation.zip*

The continuity of the tangential components of the electric field and the magnetic field leads to a recurrence relationship for the quantity $\rho_{j-1,j}$:

$$\rho_{j-1,j} = a_{j-1}^4 \frac{\rho_{j,j+1} - r_{j-1,j}}{1 + \rho_{j,j+1} r_{j-1,j}} \quad [2.97]$$

where the quantity $\rho_{j-1,j}$ is given by:

$$\rho_{j-1,j} = a_{j-1} \left(\frac{E_{j-1}^r}{E_{j-1}^f} \right) \quad [2.98]$$

where E_{j-1}^f is the amplitude of the incident electric field at the interface *j-1,j*, whereas E_{j-1}^r is the amplitude of the reflected field; a_j is the amplitude factor at half-thickness:

$$a_j = exp\left(i\, k_{zj}\, \frac{d_j}{2}\right) \qquad [2.99]$$

and lastly $r_{j-1,j}$ is the Fresnel coefficient *j–1,j*. The calculation method consists of starting with the substrate *N*, setting $\rho_{n-1,N} = 0$ and ascending towards the incident medium *1* for which $a_0 = 1$. The reflectance can be deduced using the formula $|\rho_{0,1}|^2$. The method can also be used to determine the internal distribution of the fields.

2.2.4.2. *Recursive method of MacLeod and Berning*

The recurrence relationship [2.97] can be rewritten using the quantity $\tilde{R}_{j,j+1}$ as:

$$\tilde{R}_{j-1,j} = \frac{r_{j-1,j} + exp(-2\, i\, k_{zj-1}\, d_{j-1})\tilde{R}_{j,j+1}}{1 + \tilde{R}_{j,j+1} r_{j-1,j}\; exp(-2\, i\, k_{zj-1}\, d_{j-1})}$$

$$\tilde{R}_{j-1,j} = a_{j-1}^{-4}\, \rho_{j-1,j};\, \tilde{R}_{j,j+1} = a_{j-1}^{4}\, \rho_{j,j+1} \qquad [2.100]$$

This relationship can be associated with the following recursive formula (MacLeod 1986):

$$R'_p = \frac{r_p + exp(-i\, \delta_p)R'_{p-1}}{1 + R'_{p-1}\, r_i\; exp(-i\, \delta_p)};\; \delta_p = 2\, k_{zp} d_p \qquad [2.101]$$

where the quantities R'_p, corresponding to $\tilde{R}_{j-1,j}$, and R'_{p-1}, corresponding to $\tilde{R}_{j,j+1}$, can be interpreted respectively as the amplitude of reflection above the *p*th and *p–1*th layer, the resultant of the contribution *of all the layers located below the layer involved* (see Figure 2.5). Relationship [2.101] is also obtained by calculating the sum of the multiple reflections between the interfaces of the *p*th layer:

$$R'_p = r_P^+ + t_P^+\, R'^+_{p-1} t_P^-\; exp(-i\, \delta_p)\left[1 + r_P^-\, R'^+_{p-1}\, exp(-i\, \delta_p) + \left(r_P^-\, R'^+_{p-1}\, exp(-i\, \delta_p)\right)^2 + \cdots\right] = r_P^+ + \frac{t_P^+\, R'^+_{p-1} t_P^-\, exp(-i\, \delta_p)}{1 - exp(-i\, \delta_p) r_P^-\, R'^+_{p-1}} \qquad [2.102]$$

Since the Fresnel coefficients verify the relationships: $r_P^+ \equiv r_p = -r_P^- = -r_p$, $t_P^+ = t_P^- = t_p$ and generally $r_P^+ r_P^- + t_P^+ t_P^- = 1$ (see relationship [2.63]), then:

$$R'_p = \frac{r_p + R'^+_{p-1}\, exp(-i\, \delta_p)}{1 + exp(-i\, \delta_p) r_p\, R'^+_{p-1}};\; R'^+_{p-1} \equiv R'_{p-1} \qquad [2.103]$$

This relationship has very often been used for performance simulation and synthesis of multilayer structures, mainly in the field of long wavelengths, while Parratt's iterative formulation is more commonly applied in the field of X-ray radiation. MacLeod-type formulation [2.103] is not very convenient for synthesis because the "primed" quantities are taken in reference to the material medium above the respective interfaces. The reference with respect to a vacuum is exploited in P. Berning's approach (1963).

To perform this change, we can proceed as follows: if in [2.103] the thickness d_p tends towards 0 ($\delta_p \to 0$) and we write $R'_p \to R_p, \delta_p \to 0$ and:

$$R_p = \frac{r_p + R'_{p-1}}{1 + r_p\, R'_{p-1}} \qquad [2.104]$$

which leads to:

$$R'_{p-1} = \frac{R_{p-1} - r_p}{1 - r_p\, R_{p-1}} \qquad [2.105]$$

The "non-primed" quantities *are taken in reference to a vacuum*; expressing this in other terms, r_p is the Fresnel reflection coefficient of the material of layer p with respect to a vacuum. Introducing [2.105] into [2.103], it becomes:

$$R_p = \frac{r_p(1 - r_p\, R_{p-1}) + exp(-i\,\delta_p)(R_{p-1} - r_p)}{1 - r_p\, R_{p-1} + r_p\,(R_{p-1} - r_p)\; exp(-i\,\delta_p)} \qquad [2.106]$$

Using this last relation, the amplitude of the coefficient of reflection of the overall stack can be calculated by recurrence, by setting the initial condition $R_0 = r_0$.

In the field of X-ray radiation and aside from the angular domain of the total reflection $|r_p| \ll 1$, relationship [2.106] can be rewritten in the following form, which is approximate but largely applicable in the aforementioned conditions:

$$R_p = r_p(1 - exp(-i\,\delta_p)) + R_{p-1}\, exp(-i\,\delta_p) \qquad [2.107]$$

In addition, in the field of X-ray radiation, the quantities r_p, R_p, δ_p are complex numbers ($\in \mathbb{Z}$), because the media are absorbing and r_p, R_p can be assimilated to the vectors of a two-dimensional vectorial space:

$$r_p = \lfloor r_p \rfloor\, exp(i\,\varrho_p);\; R_p = \lfloor R_p \rfloor\, exp(i\,\vartheta_p) \qquad [2.108]$$

The formula [2.107] lends itself to a vectorial graphical approach similar to the Fresnel diagram approach used in wave-like optics and in electricity. The term $exp(-i\,\delta_p)$ acts on the vectors r_p, R_p by a rotation of an angle equal to $Re[-\delta_p]$ associated with homothety with ratio $exp(-|Im[\delta_p(d_p)]|)$. From equation [2.107], it follows that if we consider the change in R_p as the thickness of layer p, denoted as

d_p, changes, then the vector $R_p(d_p)$ turns in the complex plane by an angle $Re[-\delta_p(d_p)]$ starting at the value $d_p = 0$, and undergoes a homothetic reduction with factor $exp(-|Im[\delta_p(d_p)]|)$.

If a material **a** is deposited on a substrate, when the thickness $d_\mathbf{a}$ of this material increases, the summit of the vector $R_p(d_\mathbf{a})$ performs a spiral in the complex plane around a point denoted as $\boldsymbol{r}_\mathbf{a}$ corresponding to the Fresnel reflection coefficient of material **a** with respect to a vacuum. On this spiral, a point denoted as $\boldsymbol{R}_\mathbf{a}$ appears where the reflectance (to a given wavelength) is maximal. To increase the reflectance of the structure starting from $\boldsymbol{R}_\mathbf{a}$, it is then necessary to deposit another material **b** with suitable optical characteristics starting from a so-called switching point. The curve $R_p(d_\mathbf{b})$ also spirals around a point denoted as $\boldsymbol{r}_\mathbf{b}$ passing through a point denoted as $\boldsymbol{R}_\mathbf{b}$ where the reflectance of the structure reaches a maximum reflectance. This change is shown in Figure 2.6. A detailed study of this process is given by Yamamoto and Namioka (1992). On the basis of this principle, optimized Bragg mirror deposit systems have been designed (Chauvineau et al. 1984).

2.2.4.3. *Method based on Einstein's velocity addition law*

This approach was developed by J.-M. Vigoureux in a series of articles (Vigoureux 1992). The idea is to use the generalization of Einstein's velocity addition law used in the theory of special relativity. The approach can be seen as a reformulation of the iterative methods presented in the previous sections. In its generalized form, this law is written in the following form, using the operation denoted by $\circ$:

$$v_1 \circ v_2 = \frac{v_1 + v_2}{1 + \bar{v}_1\, v_2} \qquad [2.109]$$

where v_1 and v_2 are *complex* quantities and the symbol $\bar{\ }$ denotes a type of complex conjugation to be specified. If the quantities v_1 and v_2 are real, the law of addition of parallel velocities recurs in special relativity. In the context of the optics of stratified media, it is well-known that for a system with two interfaces (therefore three media that we will denote *1*, *2* and *3*), the reflection coefficient $\mathcal{R}_2$ (the index denotes the number of diopters) is obtained from the Fresnel reflection coefficient for each diopter $r_{1,2}$, and $r_{2,3}$ using the formula:

$$\mathcal{R}_2 = \frac{r_{1,2} + r_{2,3}\, exp(-2\, i\, \phi_2)}{1 + r_{1,2}\, r_{2,3}\, exp(-2\, i\, \phi_2)}\, exp(-2\, i\, \phi_1) \qquad [2.110]$$

where ϕ_j is the optical phase shift.

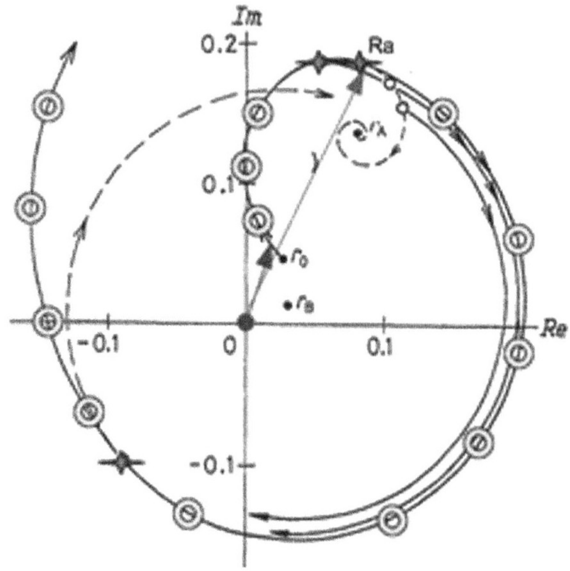

Figure 2.6. *Vectorial graphical representation in the complex plane* $[Re(R_p), Im(R_p)]$ *of change in the reflection amplitude coefficient* R_p *of the stack, as a function of the thickness of the layers and the addition of layers. For a color version of this figure, see www.iste.co.uk/andre/xrayradiation.zip*

COMMENT ON FIGURE 2.6.– *Solid line curve marked with yellow rings; the blue arrow represents a vector associated with a quantity* R_p*, the arrow in red represents a vector associated with an extremum* $\boldsymbol{R_a}$*; the red star corresponds to an extremum* $\boldsymbol{R_a}$ *and the green stars to tipping points.*

The previous formula can be rewritten as:

$$\mathcal{R}_2 = R_1 \circ R_2 = \frac{R_1 + R_2}{1 + \overline{R_1}\, R_2} \quad [2.111]$$

with:

$$R_j = r_{j,j+1}\, exp\left(-2\, i\, \textstyle\sum_{s=1}^{j} \phi_s\right);\ \bar{R}_j = r_{j,j+1}\, exp\left(+2\, i\, \textstyle\sum_{s=1}^{j} \phi_s\right) \quad [2.112]$$

We can note that the conjugation denoted by $\bar{\ }$ is not the classic complex conjugation but is instead the change of sign before the cumulative phase change. The relationship can be extended to n diopters, and we obtain:

$$\mathcal{R}_n \equiv \mathcal{R}_{(1,\dots,n)} = R_1 \circ [R_2(\circ \dots \circ R_n)] = \frac{R_1 + \mathcal{R}_{(2,\dots,n)}}{1 + \overline{R_1}\,\mathcal{R}_{(2,\dots,n)}} \quad [2.113]$$

Using the iterative form of the previous relation, an effective algorithm can be written.

2.2.4.4. *Method of impedance and admittance*

This method shows a narrow analogy with that used in the theory of electrical circuits. The impedance Z is defined by the ratio of the amplitude of the tangential component of the electric field to the amplitude of the magnetic field, where the admittance A is the inverse quantity (Z^{-1}). The impedance method is used preferably to treat the case of *TE* polarization and the admittance method is used to deal with the case of *TM* polarization.

From the expression for the Fresnel reflection coefficient, it follows that:

$$r = \frac{\xi Z - 1}{\xi Z + 1} = \frac{\xi - A}{\xi + A} \quad [2.114]$$

Starting from the method of characteristic matrices for a stratified medium (Abelès method), we deduce the recurrence formulae for the impedance Z_j and admittance A_j at the interface $j, j+1$:

$$Z_j = \frac{Z_{j-1}\cos(k_{zj}d_j) + \frac{i}{\xi_j}\sin(k_{zj}d_j)}{i\,\xi_j Z_{j-1}\sin(k_{zj}d_j) + \cos(k_{zj}d_j)} \quad [2.115]$$

$$A_j = \frac{A_{j-1}\cos(k_{zj}d_j) + i\,\xi_j\sin(k_{zj}d_j)}{\frac{i}{\xi_j}A_{j-1}\sin(k_{zj}d_j) + \cos(k_{zj}d_j)} \quad [2.116]$$

The calculation continues in a recurrent manner starting from the initial conditions:

$$Z_n = \frac{1}{\xi_n}, A_n = \xi_n \quad [2.117]$$

In addition, the impedance and admittance obey the following differential equations for *TE* polarization:

$$\frac{dZ}{dz} = i\,k[1 - (\varepsilon(z) - \alpha^2)\,Z^2]$$

$$\frac{dA}{dz} = i\,k[(\varepsilon(z) - \alpha^2) - A^2] \quad [2.118]$$

and the following equations for *TM* polarization:

$$\frac{dZ}{dz} = i\,k\left[1 - \frac{\alpha^2}{\varepsilon(z)} - \varepsilon(z)\,Z^2\right]$$

$$\frac{dA}{dz} = i\,k\left[\varepsilon(z) - \left(1 - \frac{\alpha^2}{\varepsilon(z)}\right)A^2\right] \qquad [2.119]$$

with:

$$\alpha^2 = {k_\parallel}^2/k^2 \qquad [2.120]$$

These differential equations are Riccati equations; their solution is of interest for the case of non-stratified systems, in particular for sinusoidal profiles (rough filters).

Let us note that the impedance can also be connected to the components of the matrix $\hat{C}_{unit}$ by the following relationship, obtained by combining the matrix expression for propagation within a unit cell with the continuity of the impedance, so:

$$Z = \frac{\hat{C}_{unit\,11} - \hat{C}_{unit\,22} \pm\sqrt{(\hat{C}_{unit\,11} - \hat{C}_{unit\,22})^2 + 4\hat{C}_{unit\,12}\,\hat{C}_{unit\,21}}}{2\,\hat{C}_{unit\,21}} \qquad [2.121]$$

Using this relationship and formula [2.114], we can calculate the reflection coefficient. The choice of the sign in relation [2.115] depends on the physical conditions. Therefore, for passive media, the real part of the impedance must be positive. Analogous relationships can be found using the admittance.

2.2.4.5. *Recursive method via S matrices*

Let us consider a stratified system with *M* layers characterized by the following scattering matrix (S matrix):

$$\hat{\Sigma}^{(M)} = \begin{pmatrix} \Sigma_{11}^{(M)} & \Sigma_{12}^{(M)} \\ \Sigma_{21}^{(M)} & \Sigma_{22}^{(M)} \end{pmatrix} \qquad [2.122]$$

The reflection coefficient at the interface *M*+*1* will be given by the recurrence relationship:

$$r^{(M+1)} = \Sigma_{11}^{(M+1)} + \frac{\Sigma_{12}^{(M+1)}\,\Sigma_{21}^{(M+1)} r^{(M)}\,exp(i\,\delta_M)}{1 - \Sigma_{22}^{(M+1)} r^{(M)} exp(i\,\delta_M)} \qquad [2.123]$$

with $r^{(1)} = \Sigma_{11}^{(1)}$ and where δ_j is the phase shift in the layer *j*.

2.3. Superluminal propagation and density of optical modes

The group velocity v_g describes the speed of propagation of a pulse of light through an optical structure. It has been established both theoretically and experimentally that within the stopband of a 1D-PC, v_g could exceed the speed of light in a vacuum several times, which corresponds to a so-called superluminal propagation (Spielmann et al. 1994; Pereyra and Simanjuntak 2007). In fact, within the stopband, the standard concept of group velocity loses its pertinence and a quantity $\bar{v}_g = \frac{L_W}{\tau_W}$, where L_W denotes the distance traveled within the optical structure during the phase time (or Wigner time) defined by:

$$\tau_W = \frac{d\varphi}{d\omega} \quad [2.124]$$

where φ denotes the phase and ω the frequency.

This concept is drawn from quantum mechanics in the study of the tunnel effect of an electron through a potential barrier which has resulted in the so-called Hartman paradox (1962). This author has established that the phase time is independent of the length of the barrier (under certain conditions) such that the group velocity increases linearly with L. Following the works of Bendickson et al. (1996), the group velocity can be formulated analytically as the inverse of a quantity ρ called the density of modes and denoted as ρ_N in the case of a one-dimensional photonic structure with N layers. The idea is to start with the relation that connects the density of modes ρ_N to the transmission coefficient for the structure $t_N = x_N + i\, y_N$ with:

$$x_N = \frac{x_1 \sin(N\beta)\sin(\beta) - (x_1{}^2 + y_1{}^2)\sin((N-1)\,\beta)\sin(\beta)}{\sin^2(N\beta) - 2\,x_1 \sin(N\beta)\sin((N-1)\,\beta) + (x_1{}^2+y_1{}^2)\sin^2((N-1)\,\beta)}$$

$$y_N = \frac{y_1 \sin(N\beta)\sin(\beta)}{\sin^2(N\beta) - 2\,x_1 \sin(N\beta)\sin((N-1)\,\beta) + (x_1{}^2+y_1{}^2)\sin^2((N-1)\,\beta)} \quad [2.125]$$

where:

$$\cos(\beta) = Re\left(\frac{1}{t_1}\right) \quad [2.126]$$

where β is the Bloch phase. The density of mode ρ_N is then written as:

$$\rho_N = \frac{1}{L}\frac{dz_N}{d\omega}\frac{1}{1+z_N{}^2} \quad [2.127]$$

with:

$$z_N \equiv \frac{y_N}{x_N} = z_1\, tan(N\beta)\, cotan(\beta) \qquad [2.128]$$

The maxima of ρ_N correspond to $\beta = m\frac{\pi}{N}, m = 0, \ldots, N-1$ and take as their approximate value calculated in the middle of the stopband:

$$\rho_{Nmax} \approx -\frac{1}{L}\frac{d\eta}{d\omega}\frac{sinh(\Theta)}{\eta^2 + (sinh(\Theta))^2} \qquad [2.129]$$

with:

$$\eta = \frac{y_1}{t_1}, \Theta = \frac{1}{cosh(-cos(\beta))} \qquad [2.130]$$

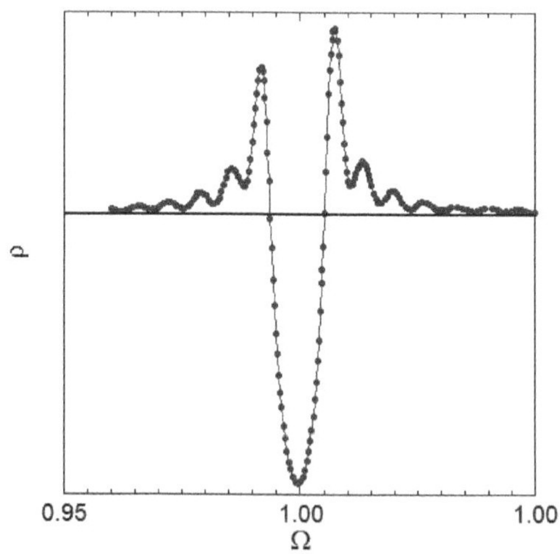

Figure 2.7. *Typical variation of the density of optical modes ρ for an absorbing system as a function of the relative frequency $\Omega = \frac{\omega}{\omega_c}$, in a Bragg diffraction regime calculated according to the method of Bendickson et al. (1996)*

COMMENT ON FIGURE 2.7.– *ω is the frequency of the radiation and ω_c is the frequency of the center of the stopband; the red line corresponding to the unit value of ρ gives the limit of the propagation regime: below this line, the superluminal regime occurs inside the stopband.*

Figure 2.7 shows the typical variation of the density of optical modes for a periodic multilayer structure near a stopband, obtained using the method described above.

2.4. References

Abelès, F. (1950a). La théorie générale des couches minces. *Journal de Physique et le Radium*, 11(7), 307–309.

Abelès, F. (1950b). Recherches sur la propagation des ondes électromagnétiques sinusoidales dans les milieux stratifiés : application aux couches minces. *Annales de Physique*, 5, 596–640.

André, J.-M. and Barchewitz, R. (1980). Specular reflection of X-rays in the anomalous dispersion region of the silicon K absorption edge. *Optics Letters*, 5(12), 543–545.

André, J.-M. and Pardo, B. (1988). Onde électromagnétique sinusoidale dans un milieu périodiquement stratifié. *Optics Communications*, 66(5/6), 249–254.

André, J.-M., Maquet, A., Barchewitz, R. (1982). X-ray specular reflectivity and anomalous scattering in the vicinity of the Si K absorption edge in quartz. *Physical Review B*, 25(9), 5671–5679.

André, J.-M., Jonnard, P., Michaelsen, C., Wiesmann, J., Bridou, F., Ravet-Krill, M.-F., Arnaud, J., Delmotte, F., Filatova, E.O. (2005). La/B4C small period multilayer interferential mirror for the analysis of boron. *X-Ray Spectrometry*, 34(3), 203.

Barchewitz, R., Cremonese-Visicato, M., Onori, G. (1978). X-ray photoabsorption of solids by specular reflection. *Journal of Physics C: Solid State Physics*, 11(21), 4439.

Bendickson, J.M., Dowling, J.P., Scalora, M. (1996). Analytic expressions for the electromagnetic mode density in finite, one-dimensional, photonic band-gap structures. *Physical Review E*, 53(4), 4107–4121.

Berning, P.H. (1963). Theory and calculations of optical thin films. In *Physics of Thin Films: Advances in Research and Development*, Hass, G. (ed.). Academic Press, New York/London, 69–122 [Online]. Available at: https://archive.org/details/PhysicsOfThinFilmsVol1.

Chauvineau, J.P., Corno, J., Naccache, D., Nevot, L., Pardo, B., Valiergue, L. (1984). The fabrication of multilayers for soft X-rays and extreme ultra-violet mirrors. *Journal of Optics*, 15(4B), 265.

Cromer, D.T. and Liberman, D. (1970). Relativistic calculation of anomalous scattering factors for X-rays. *The Journal of Chemical Physics*, 53(5), 1891–1898.

Feng, Y., Qi, R., Jiang, L., Huang, Q., Li, T., Liu, G., Li, W., Yan, W., Zhang, Z., Wang, Z. (2021). Chemical modification of B4C films and B4C/Pd layers stored in different environments. *Materials*, 14(5), 1319.

Floquet, G. (1883). Sur les équations différentielles linéaires à coefficients périodiques. *Annales Scientifiques de l'École Normale Supérieure*, 12, 47–88.

Hartman, T.E. (1962). Tunneling of a wave packet. *Journal of Applied Physics*, 33(12), 3427–3433.

Herpin, A. (1947). Calcul du pouvoir réflecteur d'un système stratifié quelconque. *Comptes Rendus Académie des Sciences Paris*, 225, 182–183.

Hill, G.W. (1886). On the part of the motion of the lunar perigee which is a function of the mean motions of the sun and moon. *Acta Mathematica*, 8(1), 1–36.

Kaplan, A.E. and Shkolnikov, P.L. (2005). Radiation efficiency of water-window Cherenkov sources using atomic-shell resonances. *Applied Physics Letters*, 86(2), 024107.

Katsidis, C.C. and Siapkas, D.I. (2002). General transfer-matrix method for optical multilayer systems with coherent, partially coherent, and incoherent interference. *Applied Optics*, 41(19), 3978–3987.

Kefi, M., André, J.-M., Héno, Y., Giorgi, G., Bonnelle, C. (1992). X-ray attenuation and anomalous scattering factors of medium-Z elements around their K edge. *Physical Review A*, 45(5), 2859–2867.

Knulst, W., Luiten, O.J., van der Wiel, M.J., Verhoeven, J. (2001). Observation of narrow-band Si L-edge Čerenkov radiation generated by 5 MeV electrons. *Applied Physics Letters*, 79(18), 2999–3001.

Kramers, H.A. (1927). La diffusion de la lumiere par les atomes. *Atti Cong. Intern. Fisica (Transactions of Volta Centenary Congress) Como*, 2, 545–557.

de Laer Kronig, R. (1926). On the theory of dispersion of X-rays. *JOSA*, 12(6), 547–557.

de Laer Kronig, R. (1931). Zur Theorie der Feinstruktur in den Röntgenabsorptionsspektren. *Zeitschrift für Physik*, 70(5), 317–323.

Lekner, J. (1994). Light in periodically stratified media. *JOSA A*, 11(11), 2892–2899.

Li, L. (1996). Formulation and comparison of two recursive matrix algorithms for modeling layered diffraction gratings. *JOSA A*, 13(5), 1024–1035.

Lytle, F.W. (1999). The EXAFS family tree: A personal history of the development of extended X-ray absorption fine structure. *Journal of Synchrotron Radiation*, 6(3), 123–134.

Macleod, H.A. (1986). *Thin-Film Optical Filters*. Macmillan, New York.

Marmoret, R. and André, J.-M. (1983). Bragg reflectivity of layered synthetic microstructures in the x-ray anomalous scattering regions. *Applied Optics*, 22(1), 17–19.

Ménesguen, Y., Lépy, M.-C., Beckhoff, B. (2018). Trends in X-ray techniques. *Spectroscopy*, 33(7), 24–38.

Montiel, F. and Nevière, M. (1994). Differential theory of gratings: Extension to deep gratings of arbitrary profile and permittivity through the R-matrix propagation algorithm. *JOSA A*, 11(12), 3241–3250.

Palik, E.D. (1985). *Handbook of Optical Constants of Solids*. Academic Press, Boston [Online]. Available at: http://www.sciencedirect.com/science/article/pii/B9780080547213500010 [Accessed on 31 August 2016].

Parajuli, R.K., Sakai, M., Parajuli, R., Tashiro, M. (2022). Development and applications of Compton camera—A review. *Sensors*, 22(19), 7374.

Parratt, L.G. (1954). Surface studies of solids by total reflection of X-rays. *Physical Review*, 95(2), 359–369.

Parratt, L.G. and Hempstead, C.F. (1954). Anomalous dispersion and scattering of X-rays. *Physical Review*, 94(6), 1593–1600.

Pereyra, P. and Simanjuntak, H.P. (2007). Time evolution of electromagnetic wave packets through superlattices: Evidence for superluminal velocities. *Physical Review E*, 75(5), 056604.

Rosenbluth, A.E. and Lee, P. (1982). Bragg condition in absorbing X-ray multilayers. *Applied Physics Letters*, 40(6), 466–468.

Rumpf, R.C. (2011). Improved formulation of scattering matrices for semi-analytical methods that is consistent with convention. *Progress in Electromagnetics Research*, 35, 241–261.

Spielmann, Ch., Szipöcs, R., Stingl, A., Krausz, F. (1994). Tunneling of optical pulses through photonic band gaps. *Physical Review Letters*, 73(17), 2308–2311.

Troparevsky, M.C., Sabau, A.S., Lupini, A.R., Zhang, Z. (2010). Transfer-matrix formalism for the calculation of optical response in multilayer systems: From coherent to incoherent interference. *Optics Express*, 18(24), 24715–24721.

Vezzetti, D.J. and Cahay, M.M. (1986). Transmission resonances in finite, repeated structures. *Journal of Physics D: Applied Physics*, 19(4), L53.

Vigoureux, J.M. (1992). Use of Einstein's addition law in studies of reflection by stratified planar structures. *JOSA A*, 9(8), 1313–1319.

Yamamoto, M. and Namioka, T. (1992). Layer-by-layer design method for soft-X-ray multilayers. *Applied Optics*, 31(10), 1622–1630.

3

Coupled-wave Theories

Jean-Michel ANDRÉ and Philippe JONNARD
LCPMR, CNRS, Sorbonne Université, Paris, France

3.1. Introduction

The coupled-wave approach originates from the work of Kogelnik (1969) on treating diffraction using holograms presenting low spatial modulation. In the case of artificial Bragg structures, this hypothesis is no longer truly verified because the optical index modulation is generally significant. However, the standard coupled-wave method is in fact applicable to the condition of remaining in the vicinity of the Bragg resonance. More elaborate coupled-wave theories (Morozov et al. 2004a, 2004b) exist than the theory presented here, with some even qualified as "rigorous" (Moharam and Gaylord 1981), but these require significant numerical calculations. Nevertheless, using the standard approach developed hereafter, we can understand the physical processes modeled in an overall and satisfactory manner. This approach also has the advantage that it deals with propagation in Bragg structures that we qualify as hybrid. These are structures constructed from or using Bragg mirrors.

In this chapter, we also consider the approach known as the Tagaki–Taupin (T–T) approach after the name of the two scientists who developed it independently to deal with diffraction of X-ray radiation by distorted crystals, involving dislocations (Taupin 1964, 1967). The approach was extended by looking at the case of quasi-periodic deformed structures; by thermal effects, for example. In terms of mathematics, this approach comes down to a system of coupled differential equations, which explains why it is considered in this chapter. In its usual form, it uses the two-wave approximation but it has been extended to a system of N waves to deal

X-ray Radiation and Artificial Bragg Structures,
coordinated by Jean-Michel ANDRÉ and Philippe JONNARD.

with the problems of diffraction in volume of different optical components, in particular the Fresnel zones (Yan et al. 2007). Using this model, the case of imperfections affecting the multilayer structures in the stratification plane can be examined.

The coupled-wave approach also allows the problem of propagation of a “luminous” pulse with a finite time width (so not described by a monochromatic plane wave) to be dealt with directly. This problem is generally treated using a Fourier time transform: the Green impulse response corresponding to a Dirac $\delta(t)$ incident pulse is calculated, allowing the response of an incident pulse with an arbitrary time profile to be determined using the convolution theorem.

In physical terms, the configured problem results primarily from the fact that for a short source which consequently presents a broad spectrum – which means that the pulse is made up of many superimposed frequencies – these propagate at different velocities (spectral dispersion). The design of chirped mirrors, developed to compensate for this spectral dispersion, is most often based on a Fourier-type mathematical approach; the subject has been studied by Bourassin-Bouchet et al. (2011) for ultrashort EUV pulses.

However, analysis in the time domain (and not the frequency domain) is more direct and can be performed without major mathematical difficulties using the coupled-wave theory. This type of approach was firstly considered in the context of crystallography, in particular by Chukhovskii and Förster (1995), then extended to the optical domain by André and Jonnard (2015).

3.2. Frequency domain

3.2.1. *Standard theory of coupled waves*

From a mathematical point of view, the principle of the method resembles the constant variation method (Lagrange method) to resolve differential equations with variable coefficients. The solution to the propagation equation (for the electric field) is thus written in the form:

$$E(z) = \tilde{\mathcal{F}}(z) exp(i\, k_z\, z) + \tilde{\mathcal{R}}(z)\, exp(-i\, k_z\, z) \quad [3.1]$$

where the “constants” $\tilde{\mathcal{F}}$ and $\tilde{\mathcal{R}}$ occur in the solution of a homogeneous environment, becoming functions of the space variable z. Usually, the dielectric constant $\varepsilon(z)$ is expanded into a Fourier series:

$$\varepsilon(z) = \bar{\varepsilon} + \sum_{n=-\infty}^{+\infty} \varepsilon_n \, exp(i\, n\, g\, z)\,; n \in \mathbb{Z}^* \qquad [3.2]$$

or in terms of the dielectric susceptibility $\chi(z)$:

$$\chi(z) \equiv \frac{\varepsilon(z)-1}{4\pi} = \left(\bar{\chi} + \sum_{n=-\infty}^{+\infty} \chi_n \, exp(i\, n\, g\, z)\right), n \in \mathbb{Z}^* \qquad [3.3]$$

If we consider that the structure is a stratified system of bilayers with a division parameter γ and a susceptibility contrast $\Delta\chi$:

$$\varepsilon_n = 4\pi\, \chi_n \,;\, \chi_n = u_n\, \Delta\chi\,;\, u_n = -\frac{i}{2n\pi}\left[1 - exp(-\, i\, 2\, \pi\, n\, \gamma)\right] \qquad [3.4]$$

Introducing these expressions into the propagation equation, the following differential equation is obtained:

$$\left[\frac{d^2\tilde{\mathcal{F}}(z)}{dz^2} + 2\, i\, \kappa\, \frac{d\tilde{\mathcal{F}}(z)}{dz} + k^2\, \tilde{\mathcal{F}}(z) \sum_{n=-\infty}^{+\infty} \varepsilon_n \, exp(i\, n\, g\, z)\right]\left[exp\left\{z\left(i\, \tilde{n}\, \frac{g}{2} - \frac{\alpha - 2\, i\, \delta}{2}\right)\right\}\right] +$$
$$\left[\frac{d^2\tilde{\mathcal{R}}(z)}{dz^2} - 2\, \kappa\, \sqrt{\bar{\varepsilon}}\, \frac{d\tilde{\mathcal{R}}(z)}{dz} + k^2\, \tilde{\mathcal{R}}(z) \sum_{n=-\infty}^{+\infty} \varepsilon_n \, exp(i\, n\, g\, z)\right]\left[exp\left\{z\left(-i\, \tilde{n}\, \frac{g}{2} + \frac{\alpha - 2\, i\, \delta}{2}\right)\right\}\right] = 0 \qquad [3.5]$$

where δ is the detuning coefficient given by:

$$\delta = \kappa - \tilde{n}\, \frac{g}{2} \qquad [3.6]$$

This quantity is cancelled out when Bragg resonance occurs at order $\tilde{n}$. In the above, the term κ denotes $k_z \sqrt{Re(\bar{\varepsilon})}$ and the term α is written as:

$$\alpha = \frac{k\, Im(\bar{\varepsilon})}{\sqrt{Re(\bar{\varepsilon})}} \qquad [3.7]$$

Using this last equation, a potential phenomenon of absorption or gain can be introduced.

Following Kogelnik's hypothesis, which states that the exchange between the direct wave and the retrograde wave is weak along a spatial period, then the functions $\tilde{\mathcal{F}}(z)$ and $\tilde{\mathcal{R}}(z)$ and their derivatives, and similarly the exponential factors between $\lfloor \ldots \rfloor$, can be considered constants over a period. It follows that after an "averaging" operation over a period, a system of coupled equations is obtained, which approximately connect the amplitudes $\tilde{\mathcal{F}}(z)$ and $\tilde{\mathcal{R}}(z)$ as follows:

$$\frac{d^2\tilde{\mathcal{F}}(z)}{dz^2} + 2\,i\,\kappa\,\frac{d\tilde{\mathcal{F}}(z)}{dz} + k^2\,\tilde{\mathcal{R}}(z)\,\varepsilon_{\bar{n}}\,e^{z(\alpha-2\,i\,\delta)} + 2\,\kappa\left(2\pi\,\frac{k^2}{\kappa}\,\bar{\chi} - \delta\right)\tilde{\mathcal{F}}(z) = 0$$

$$\frac{d^2\tilde{\mathcal{R}}(z)}{dz^2} - 2\,i\,\kappa\,\frac{d\tilde{\mathcal{R}}(z)}{dz} + k^2\,\tilde{\mathcal{F}}(z)\,\varepsilon_{-\bar{n}}e^{-z(\alpha-2\,i\,\delta)} + 2\,\kappa\left(2\pi\,\frac{k^2}{\kappa}\,\bar{\chi} - \delta\right)\tilde{\mathcal{R}}(z) = 0 \qquad [3.8]$$

The conditions that must be met for the quantities listed above to be effectively considered constants are the following:

$$\alpha d \ll 1, |\,\delta|\,d \ll 1, \left|\frac{k\,\varepsilon_{\pm\,\bar{n}}}{2\,\sqrt{\bar{\varepsilon}}}\right|\,d \ll 1 \qquad [3.9]$$

If we place ourselves in the approximation of *slowly varying amplitude* (SVA), the second derivatives are neglected and the amplitudes $\tilde{\mathcal{R}}(z)$ and $\tilde{\mathcal{F}}(z)$ are connected by a system of first-order coupled differential equations that can be written in the following matrix form:

$$\frac{d}{dz}\begin{pmatrix}\tilde{\mathcal{F}}(z)\\ \tilde{\mathcal{R}}(z)\end{pmatrix} = i\,\bar{\mathcal{M}}\begin{pmatrix}\tilde{\mathcal{F}}(z)\\ \tilde{\mathcal{R}}(z)\end{pmatrix} \qquad [3.10]$$

with:

$$\bar{\mathcal{M}} = \begin{pmatrix}-\tilde{\alpha} & \widetilde{K^{+}}\\ \widetilde{K^{-}} & \tilde{\alpha}\end{pmatrix}$$

$$\widetilde{K^{\pm}} = \mp 2\pi\,\frac{k^2}{\kappa}\,\chi_{\bar{n}}$$

$$\tilde{\alpha} = 2\pi\,\frac{k^2}{\kappa}\,\bar{\chi} - \delta \qquad [3.11]$$

The problem of coupled waves can be related to an eigenvalues problem. We write:

$$\begin{pmatrix}\tilde{\mathcal{F}}(z)\\ \tilde{\mathcal{R}}(z)\end{pmatrix} = \begin{pmatrix}\tilde{\mathcal{F}}\\ \tilde{\mathcal{R}}\end{pmatrix}e^{i\,\tilde{\lambda}\,z} \qquad [3.12]$$

such that the following eigenvalue equation is obtained:

$$\left(\bar{\mathcal{M}} - \tilde{\lambda}\,\mathbb{I}_2\right)\begin{pmatrix}\tilde{\mathcal{F}}\\ \tilde{\mathcal{R}}\end{pmatrix} = 0 \qquad [3.13]$$

where $\mathbb{I}_2$ is the identity matrix with dimension 2. The solutions $\tilde{\lambda}^{\pm}$ to [3.13] are given by:

$$\tilde{\lambda}^{\pm} = \pm\sqrt{\tilde{\alpha}^2 + \widetilde{K^{-}}\widetilde{K^{+}}} \qquad [3.14]$$

Supposing that $\tilde{\alpha}$ and $\widetilde{K^{\pm}}$ are purely real quantities, in other words, for the case of non-absorbing and no-gain environments, then $\widetilde{K^{+}} = -\widetilde{K}$ and $\widetilde{K^{-}} = +\widetilde{K}$ with $\widetilde{K} = 2\pi \frac{k^2}{\kappa} \chi_{\tilde{n}}$, the quantity $\sqrt{\tilde{\alpha}^2 + \widetilde{K^{-}}\widetilde{K^{+}}} = \sqrt{\tilde{\alpha}^2 - \widetilde{K}^2}$ will then be purely imaginary if $|\tilde{\alpha}| \leq |\widetilde{K}|$, which corresponds to a stopband for a contra-directional coupling. The width of the stopband in terms of frequency $\Delta\omega$ for an infinite structure is given by:

$$\Delta\omega = 2\, v_g\, |\widetilde{K}| \qquad [3.15]$$

where v_g is the group velocity for the mode considered. The eigenvectors $\boldsymbol{V}^{\pm}$ associated with $\tilde{\lambda}^{\pm}$ respectively are given by:

$$\boldsymbol{V}^{+} = \begin{pmatrix} \widetilde{K^{+}} \\ \tilde{\alpha} + \sqrt{\tilde{\alpha}^2 - \widetilde{K}^2} \end{pmatrix} ; \boldsymbol{V}^{-} = \begin{pmatrix} \widetilde{K^{+}} \\ \tilde{\alpha} - \sqrt{\tilde{\alpha}^2 - \widetilde{K}^2} \end{pmatrix} \qquad [3.16]$$

Change in $\begin{pmatrix} \tilde{\mathcal{F}}(z) \\ \tilde{\mathcal{R}}(z) \end{pmatrix}$ is expressed by the eigenmatrix $\overline{V} = (\boldsymbol{V}^{+}\boldsymbol{V}^{-})$ and the formula:

$$\begin{pmatrix} \tilde{\mathcal{F}}(z) \\ \tilde{\mathcal{R}}(z) \end{pmatrix} = \overline{S(z)} \begin{pmatrix} \tilde{\mathcal{F}}(0) \\ \tilde{\mathcal{R}}(0) \end{pmatrix} \qquad [3.17]$$

with:

$$\overline{S(z)} = V \begin{pmatrix} e^{i\tilde{\lambda}^{+} z} & 0 \\ 0 & e^{i\tilde{\lambda}^{-} z} \end{pmatrix} \overline{V}^{-1} = \begin{pmatrix} cos(\varsigma(z)) - i \frac{\tilde{\alpha}}{\sqrt{\tilde{\alpha}^2 - \widetilde{K}^2}} sin(\varsigma(z)) & - \frac{i\widetilde{K}}{\sqrt{\tilde{\alpha}^2 - \widetilde{K}^2}} sin(\varsigma(z)) \\ \frac{i\widetilde{K}}{\sqrt{\tilde{\alpha}^2 - \widetilde{K}^2}} sin(\varsigma(z)) & cos(\varsigma(z)) + i \frac{\tilde{\alpha}}{\sqrt{\tilde{\alpha}^2 - \widetilde{K}^2}} sin(\varsigma(z)) \end{pmatrix} \qquad [3.18]$$

where:

$$\varsigma(z) = \sqrt{\tilde{\alpha}^2 - \widetilde{K}^2}\, z \qquad [3.19]$$

Using relationships [3.17], [3.18] and [3.19], the reflection r and transmission t coefficients of a crystal with length L can be calculated:

$$r \equiv \left.\frac{\tilde{\mathcal{R}}(0)}{\tilde{\mathcal{F}}(0)}\right]_{\tilde{\mathcal{R}}(L)=0} = -\frac{S_{21}\,(L)}{S_{22}\,(L)} = \frac{i\,\tilde{K}\,sin(\varsigma(L))}{\sqrt{\tilde{\alpha}^2-\tilde{K}^2}\,cos(\varsigma(L))+i\,\tilde{\alpha}\,sin(\varsigma(L))}$$

$$t \equiv \left.\frac{\tilde{\mathcal{F}}(L)}{\tilde{\mathcal{F}}(0)}\right]_{\tilde{\mathcal{R}}(L)=0} = \frac{det\,\overline{S\,(L)}}{S_{22}\,(L)} = \frac{\sqrt{\tilde{\alpha}^2-\tilde{K}^2}}{\sqrt{\tilde{\alpha}^2-\tilde{K}^2}\,cos(\varsigma L)+i\,\tilde{\alpha}\,sin(\varsigma(L))} \quad [3.20]$$

Within a stopband, the quantity $\varsigma(z)$ becomes imaginary and we write $\varsigma(z) = i\,c(z), sin(\varsigma(z)) = i\,sinh(c(z)), cos(\varsigma(z)) = cosh(c(z))$, such that:

$$r = \frac{-\,\tilde{K}\,sinh(c(L))}{\sqrt{\tilde{\alpha}^2-\tilde{K}^2}\,cosh(c(L))-\tilde{\alpha}\,sinh(c(L))}$$

$$t = \frac{\sqrt{\tilde{\alpha}^2-\tilde{K}^2}}{\sqrt{\tilde{\alpha}^2-\tilde{K}^2}\,cosh(c(L))-\tilde{\alpha}\,sinh(c(L))} \quad [3.21]$$

At the Bragg resonance for $\tilde{\alpha} = 0$ (non-refringent, non-absorbing and passive medium), for the reflectance, we deduce the formula:

$$|r|^2 = \left(tanh\left|\tilde{K}\,L\right|\right)^2 \quad [3.22]$$

which indicates that the reflectance tends to 1 when $\left|\tilde{K}\,L\right|$ increases and cancels out when:

$$\left|\tilde{\alpha}/\tilde{K}\right| = \sqrt{1+\left(\frac{n\,\pi}{|\tilde{K}\,L|}\right)^2}\;;\; n \in \mathbb{N} \quad [3.23]$$

The standard coupled-wave method from Kogelnik's work provides a description of the propagation in an artificial Bragg crystal or a 1D-PC that is relatively precise near low-order Bragg domains (1, 2). It is recognized that an increase in the order, in the ratio of the period to the wavelength (d/λ) or in the index contrast reduces the precision of the method. This results from the increase in the effects of multiple diffraction and of the significant involvement of the wave number in the diffraction process.

The case of the lamellar grating (see Figure 3.1(a)) is analyzed in more depth in the context of coupled waves, because this type of structure has been largely developed as a monochromator with a narrow band or as a polychromator (see Chapter 5).

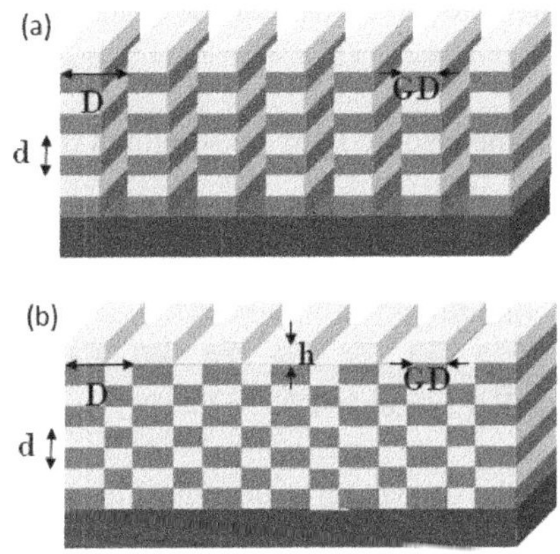

Figure 3.1. *(a) Diagram of a lamellar grating etched in a multilayer Bragg mirror formed from bilayers; (b) diagram of an alternate grating; the period of the multilayer (thickness of the bilayer) is d, the period of the grating is D and the width of the line is GD. The shift of the "column mirrors" is denoted h. For a color version of this figure, see www.iste.co.uk/andre/xrayradiation.zip*

To extend the standard theory of coupled waves to this type of structure, we start with the two-dimensional Helmholtz equation [2.9], so:

$$\nabla^2 E(x,z) + k^2\left(1 - \tilde{\chi}(x,z)\right)E(x,z) = 0 \qquad [3.24]$$

with, for *TE* polarization:

$$\tilde{\chi}(x,z) = \chi_{TE}(x,z) = U(x)\,\chi(x,z) \qquad [3.25]$$

and for *TM* polarization (Kozhevnikov et al. 2011):

$$\tilde{\chi}(x,z) = \chi_{TM}(x,z) = \chi(x,z) + \frac{1}{2k^2}\frac{\nabla^2\chi(x,z)}{1-\chi(x,z)} - \frac{3}{4\,k^2}\left(\frac{\nabla^2\chi(x,z)}{1-\chi(x,z)}\right)^2 \qquad [3.26]$$

with:

$$\chi(x,z) = \left(\bar{\chi} + \textstyle\sum_{n=-\infty}^{+\infty}\chi_n\, exp\left(i\,n\,g\,(z\cos\varphi + x\sin\varphi\,)\right)\right), n\, \in \mathbb{Z}^*$$

$$U(x) = \mathrm{G} + \sum_{t\neq 0}\frac{\left(1-exp(-2\,i\,\pi\,t\,\mathrm{G})\right)}{2\,i\,\pi\,t}\,exp\left(\frac{2\,i\,\pi\,t\,x}{D}\right) \qquad [3.27]$$

The angle φ in [3.27] is the angle between the axis perpendicular to the layers and the axis perpendicular to the external surface of the grating, in the case of inclined networks.

The solutions in terms of the electric field are sought in the form:

$$E_n(z) = \widetilde{\mathcal{F}_n}(z) exp(i\, k_{zn}\, z) + \widetilde{\mathcal{R}_n}(z)\, exp(-i\, k_{zn}\, z) \qquad [3.28]$$

with the additional unicity condition:

$$\dot{\widetilde{\mathcal{F}_n}}(z) exp(i\, k_{zn}\, z) + \dot{\widetilde{\mathcal{R}_n}}(z)\, exp(-i\, k_{zn}\, z) = 0 \qquad [3.29]$$

where the symbol $\dot{..}$ denotes the derivation with respect to the variable z.

In practice, this type of Bragg lamellar grating is used in so-called single-mode or single-order conditions, meaning that a single diffraction order is favored (Yang et al. 2015). In the application of a narrow band monochromator, developed in Chapter 7, the condition for specular reflection is implemented and only the zero order is involved in the calculation. Equation [3.10] is applied to the zero order of diffraction:

$$\frac{d}{dz}\begin{pmatrix} \widetilde{\mathcal{F}_0}(z) \\ \widetilde{\mathcal{R}_0}(z) \end{pmatrix} = i\, \overline{\mathcal{M}} \begin{pmatrix} \widetilde{\mathcal{F}_0}(z) \\ \widetilde{\mathcal{R}_0}(z) \end{pmatrix} = \overline{D} \begin{pmatrix} \widetilde{\mathcal{F}_0}(z) \\ \widetilde{\mathcal{R}_0}(z) \end{pmatrix} \qquad [3.30]$$

If instead we consider that the specular case ($n = 0$), the more general case of diffraction at an order $n \neq 0$, the matrix $\overline{D}$ from [3.30] is written as:

$$\overline{D} = \begin{pmatrix} -\frac{i\, k^2}{2\, k_{Z0}}\, \mathrm{G}\, \bar{\chi} & -\frac{i\, k^2}{2\, k_{Z0}}\, \mathrm{G}\, \Delta\chi\, u_n exp(+2\, i\, \kappa\, z) \\ \frac{i\, k^2}{2\, k_{Z0}}\, \mathrm{G}\, \Delta\chi\, u_{-n}\, exp(-2\, i\, \kappa\, z) & \frac{i\, k^2}{2\, k_{Z0}}\, \mathrm{G}\, \bar{\chi} \end{pmatrix} \qquad [3.31]$$

with:

$$\kappa = \frac{n\, \pi}{d} - k_{Z0} \qquad [3.32]$$

a quantity which at a Bragg resonance at order n cancels out ($\kappa \approx 0$).

At this stage, calculation of the reflection and transmission efficiencies continues as it did in the case of the "conventional" mirror considered previously, but instead using $\chi_{TE}(x, z)$ and $\chi_{TM}(x, z)$ given respectively by equations [3.25] and [3.26].

This coupled-wave approach constitutes a theoretical basis which also allows various cases to be treated that present as a variation on the model represented in Figure 3.1. We can cite the case of an inclined grating and the case of a grating where the line profile is not rectangular but varies with depth, meaning that the quantity G varies in this case with *z*. Various profiles have been studied in several works (André et al. 2013, 2014b). Figure 3.2 illustrates this analysis for different profiles (André et al. 2014b): rectangular, trapezoid and inverse trapezoid, in comparison with the measured diffraction pattern; the cases are specified in the key.

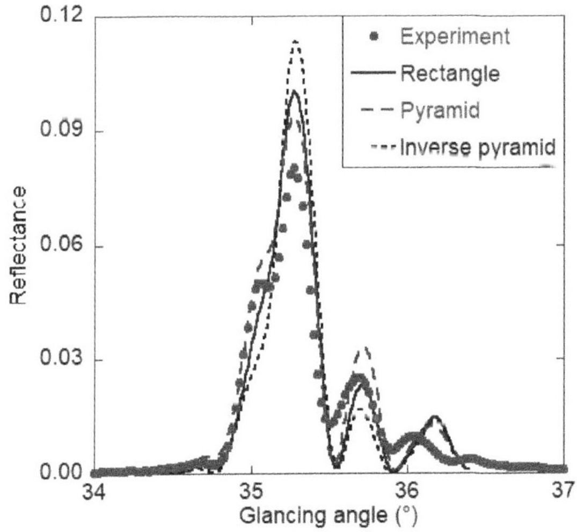

Figure 3.2. *Diffraction figure for a quasi-lamellar grating with different line profiles, for energy of 182 eV; the calculations are performed using the coupled-wave method and compared to the experimental results; the mirror is composed of 200 bilayers Mo/B_4C with thickness 6.11 nm with a division ratio equal to 0.37. For a color version of this figure, see www.iste.co.uk/andre/xrayradiation.zip*

The case of alternate gratings has also been treated with this approach (Yang et al. 2017). In this case, we write:

$$\chi(x,z) = \chi_{conv}(z)\, U(x) + \chi_{conv}(z-h)\big(1 - U(x)\big) \qquad [3.33]$$

where $\chi_{conv}(z)$ is the susceptibility of a conventional mirror given by equation [3.3] and h is the shift (see Figure 3.1(b)). One of the interesting specific aspects of this type of alternate grating is that it presents a specular reflectance of zero if both of the following conditions are met:

$$G = \frac{1}{2}, t\; h = \frac{d}{2}, t \in \mathbb{N} \qquad [3.34]$$

which are conditions corresponding to the conditions of destructive interferences between the waves diffracted by the reflective columns shifted at depth by the value h. In this zero reflectance regime, the structure, with a traversed thickness L, transmits a wave with intensity that is "simply" equal to the incident wave, attenuated according to the Beer–Lambert law with an attenuation coefficient given by $exp(-k\, L\, Im\overline{\chi}\,)$.

3.2.2. *Tagaki–Taupin method*

We consider a scalar approach where the electromagnetic field is reduced to the wave field $\psi(\boldsymbol{r})$. In the context of a two-wave approximation, this is constituted by superimposing two waves, one $\psi_i(\boldsymbol{r})$ associated with an incident wave with a wave vector $\boldsymbol{k_i}$ and the other $\psi_d(\boldsymbol{r})$ corresponding to a diffracted wave with wave vector $\boldsymbol{k_d}$:

$$\psi(\boldsymbol{r}) = \psi_i(\boldsymbol{r}) + \psi_d(\boldsymbol{r}) = \widetilde{\psi_\iota}(\boldsymbol{r})\, exp(i\, \boldsymbol{k_i}.\boldsymbol{r}) + \widetilde{\psi_d}(\boldsymbol{r})\, exp\big(i\, \boldsymbol{k_d}\, .\boldsymbol{r}\big) \qquad [3.35]$$

Inserting equation [3.35] into the Helmholtz equation (see Chapter 2) and neglecting in the Laplacian expansion $\nabla^2\psi(\boldsymbol{r})$ the $\nabla^2\widetilde{\psi_{\iota/d}}(\boldsymbol{r})$ terms which are slowly varying compared to the exponential terms, the following system of coupled differential equations (T–T equations) is obtained:

$$2\, \boldsymbol{k_i}.\vec{\nabla}\widetilde{\psi_\iota}(\boldsymbol{r}) = i\left[k^2\,(1+\chi_0) - {k_i}^2\right]\widetilde{\psi_\iota}(\boldsymbol{r}) + i\, k^2\, \chi_{\bar{1}}\, \widetilde{\psi_d}(\boldsymbol{r})\, exp(i\, \boldsymbol{g}.\boldsymbol{\partial})$$

$$2\, \boldsymbol{k_d}.\vec{\nabla}\widetilde{\psi_d}(\boldsymbol{r}) =$$
$$i\left[k^2\,(1+\chi_0) - {k_d}^2\right]\widetilde{\psi_d}(\boldsymbol{r}) + i\, k^2\, \chi_1\, \widetilde{\psi_\iota}(\boldsymbol{r})\, exp\big(-i\, \boldsymbol{g}.\boldsymbol{\partial}\big) \qquad [3.36]$$

where $\boldsymbol{g}$ is the vector of the reciprocal grating of the 1D-PC and χ_p is the pth coefficient of the Fourier expansion of the dielectric susceptibility: $\chi(\boldsymbol{r}) = \sum_p \chi_p\, exp(i\, p\, \boldsymbol{g}.\boldsymbol{r}\,)$. The vector $\boldsymbol{\partial}$ characterizes a vertical displacement (perpendicular to the stratification plane) and allows height defects in the structure layers to be taken into account. There are analytical solutions to the T–T equations in particular for "ideal" multilayer planes ($\boldsymbol{\partial} = \boldsymbol{0}$). For such structures and neglecting polarization, the system of equations T–T is written in Cartesian coordinates where z denotes the direction perpendicular to stratification and x the direction orthogonal to it:

$$(cos(\theta_B)\, \partial_x + sin(\theta_B)\, \partial_z)\, T(x,z) = i\, \frac{k}{2}\, \chi_0\, T(x,z) + i\, \frac{k}{2}\, \chi_{\bar{1}}\, R(x,z)$$

$$(cos(\theta_B)\, \partial_x - \ sin(\theta_B)\, \partial_z)\, R(x,z) =$$
$$i\, \frac{k}{2}\, (\chi_0 - \ \tilde{\alpha})\, R(x,z) + \ i\, \frac{k}{2}\, \chi_1\, T(x,z) \quad [3.37]$$

where θ_B is the Bragg angle for the wavelength $\lambda = \frac{2\,\pi}{k}$ diffracted with order 1 and $T(x,z)$ and $R(x,z)$ are the respective amplitudes of the transmitted and retrograde wave propagating under angle $\frac{\pi}{2} - \ \theta$. The quantity $\tilde{\alpha}$, which is equal to $2\, sin(2\theta_B)\, (\theta_B - \theta)$, characterizes the shift with respect to the Bragg angle. For an "ideal" and laterally unlimited structure, $\partial_x \equiv 0$, and under these conditions, $T(x,z)$ and $R(x,z)$ can be written in the form:

$$T(x,z) = C_1\, exp\left(i\, \frac{z}{sin(\theta_B)}\, \mu_1\right) + \ C_2\, exp\left(i\, \frac{z}{sin(\theta_B)}\, \mu_2\right)$$

$$R(x,z) = C_1\, \rho_1\, exp\left(i\, \frac{z}{sin(\theta_B)}\, \mu_1\right) + \ C_2\, \rho_2\, exp\left(i\, \frac{z}{sin(\theta_B)}\, \mu_2\right) \quad [3.38]$$

with:

$$\mu_{1,2} = \frac{-\,\sigma \pm \sqrt{\sigma^2 + 4\left(\sigma \frac{k}{2} \chi_0 + \left(\frac{k}{2}\right)^2 (\chi_0{}^2 - \ \chi_1 \chi_{\bar{1}})\right)}}{2}\ ;\ \sigma = \ -\ \tilde{\alpha}\ \frac{k}{2}\, ; \rho_{1,2} = \frac{\left(\mu_{1,2} - \frac{k}{2}\chi_0\right)}{\left(\chi_{\bar{1}} \frac{k}{2}\right)} \quad [3.39]$$

The coefficients C_1 and C_2 are obtained from the limit conditions. Under the usual conditions of a single incident wave in $z = 0$, the reflectance $|r_{TT}|^2$ is deduced:

$$|r_{TT}|^2 = \left|\ \rho_1 \rho_2\ \frac{exp\left(\frac{i\,L}{sin(\theta_B)}\, (\mu_2 - \mu_1)\right) - 1}{\rho_2\, exp\left(\frac{i\,L}{sin(\theta_B)}\, (\mu_2 - \mu_1)\right) - \rho_1}\right|^2 \quad [3.40]$$

For the problem of calculation of the reflectance and transmittance of a crystal with rectangular form, with limited lateral extension and a total finite thickness (of value *L*), the Laplace transform technique can be used. We summarize below this approach, which was initially developed by Kolosov and Punegov (2005). In the case considered and without taking into account the polarization, system [3.40] can be simplified into the following system:

$$\frac{\partial T}{\partial x'} + \ \frac{\partial T}{\partial z'} \ = a_1\, T + b_1\, R$$

$$\frac{\partial R}{\partial x'} - \ \frac{\partial R}{\partial z'} \ = a_2\, R + b_2\, T \quad [3.41]$$

The quantities x' and z' are reduced variables:

$$x' = \frac{x}{L}\tan(\theta_B), z' = \frac{z}{L} \quad [3.42]$$

and the coefficients a_1, a_2, b_1 and b_2 are given by the relationships:

$$a_1 = \frac{i\,k\,L}{2\,sin(\theta_B)}\chi_0\,;\; b_1 = \frac{i\,k\,L}{2\,sin(\theta_B)}\chi_{\bar{1}}, a_2 \cong \frac{i\,k\,L}{2\,sin(\theta_B)}\chi_0, b_2 = \frac{i\,k\,L}{2\,sin(\theta_B)}\chi_1 \quad [3.43]$$

The "partial" Laplace transform, along the axis of the reduced coordinate x', gives the following differential system for the transformed expressions $T(s,z')$ and $R(s,z')$:

$$\frac{\partial T(s,z')}{\partial z'} = (a_1 - s)\,T(s,z') + b_1\,R(s,z') + 1$$

$$\frac{\partial R(s,z')}{\partial z'} = -(a_2 - s)\,R(s,z') + b_2\,T(s,z') \quad [3.44]$$

with the limit conditions $(s,0) = s^{-1}$, $R(s,1) = 0$. Integration of equation [3.44] gives analytical expressions of $T(s,z')$ and $R(s,z')$; these are very long and will not be given precisely. We restrict our discussion to considering the so-called kinematic case where $b_1 = 0$. In this case, an inverse Laplace transform gives:

$$T(x',z') = exp(a_1x') - [exp(a_1x') - exp(a_1z')]\,\Theta(x'-z') \quad [3.45]$$

where Θ is the Heaviside step function. The expression of $R(x',z')$ can be put into the form:

$$R(x',z') = R_0 + R_1\,\Theta(x'-z') + R_2\,\Theta(x'+z'+1) + R_3\,\Theta(x'+z'-2)$$

$$R_0 = b_2\,\frac{exp(a_1x') - exp(a_2x')}{a_1 - a_2}$$

$$R_1 = -b_2\left[\frac{1}{a_1+a_2} + \frac{exp(a_1(x'-z'))}{a_1-a_2} - \frac{2\,a_1\,exp\big((a_1+a_2)(x'-z')/2\big)}{{a_1}^2 - {a_2}^2}\right]$$

$$R_2 = -b_2\,\frac{exp(a_2(1-z'))}{a_1-a_2}[\,exp(a_1(x'+z'-1)) - exp(a_2(x'+z'-1))]$$

$$R_3 = -b_2\,exp(a_1 + a_2(1-z'))\left[\frac{1}{a_1+a_2} + \frac{exp(a_1(x'+z'-2))}{a_1-a_2} - \frac{2\,a_1\,exp\big((a_1+a_2)(x'+z'-2)/2\big)}{{a_1}^2-{a_2}^2}\right] \quad [3.46]$$

The reflectance and transmittance are obtained by integration on x' and z'. The calculations show that the form of the diffraction pattern *depends on this lateral extension:* interferences of the waves coming from the surfaces $x' = 0$ and $z' = 0$ cause oscillations which tend to disappear when the lateral extension increases.

The problems tackled by the T–T method are generally treated solely by numerical methods, with the finite element method often being used. We finish by highlighting that the T–T method is well suited for calculating the reflectance (transmittance) of non-planar periodically stratified structures, in particular elliptically curved structures, which are important for focusing optics (Osterhoff et al. 2012). The method is also effective in tackling the diffuse scattering generated by the figure errors (correlated) or diffuse interfaces modeled by a transition layer. Finally, we point out that the T–T method is a useful tool for studying Laue diffraction using sectioned multilayer structures that can be used to create lenses or focusing devices in the X-ray domain.

3.3. Time domain

In the context of SVA approximation, the envelope $E_0(z,t)$ of the electric field of the light pulse obeys the following spatio-temporal propagation equation (time-dependent Schrödinger equation), employing the usual notations for this structure:

$$\frac{\partial^2 E_0(z,t)}{\partial z^2} + {k_z}^2(z)\, E_0(z,t) + 2\, i\, \varepsilon(z) \frac{\omega}{c^2} \frac{\partial E_0(z,t)}{\partial t} = 0 \qquad [3.47]$$

In the spirit of the coupled-wave theory, the envelope $E_0(z,t)$ is written as the superposition of two waves that propagate in opposite directions with amplitudes $\tilde{\mathcal{F}}(z,t)$ and $\tilde{\mathcal{R}}(z,t)$ dependent on the space coordinate z as well as the time coordinate t:

$$E_0(z,t) = \tilde{\mathcal{F}}(z,t) exp(i\, k_z\, z) + \tilde{\mathcal{R}}(z,t)\, exp(-i\, k_z\, z) \qquad [3.48]$$

In the vicinity of the pth Bragg resonance where $\frac{p\, g}{2} - k_z \approx 0$, the problem can be expressed using matrix formalization:

$$\frac{\partial \bar{\mathcal{E}}(z,t)}{\partial z} = \bar{\mathcal{T}}\, \bar{\mathcal{E}}(z,t) + i\, \bar{\mathcal{M}}\, \bar{\mathcal{E}}(z,t) \qquad [3.49]$$

with $\bar{\mathcal{E}}(z,t) = \begin{pmatrix} \tilde{\mathcal{F}}(z,t) \\ \tilde{\mathcal{R}}(z,t) \end{pmatrix}$ and $\bar{\mathcal{T}} = \begin{pmatrix} \frac{-k}{c\, k_z} & 0 \\ 0 & \frac{k}{c\, k_z} \end{pmatrix}$.

The solution to equation [3.49] can be expressed as follows:

$$\bar{\mathcal{E}}(z,t) = \bar{R}(z,t)\,\bar{\mathcal{E}}(0,0) \qquad [3.50]$$

with:

$$\bar{R}(z,t) = exp(-\bar{G}\,t)\,\overline{S\,(z)} \qquad [3.51]$$

where $\overline{S\,(z)}$ is given by equation [3.18] and:

$$\bar{G} = i\,\bar{\mathcal{T}}^{-1}(\psi\,\mathbb{I}_2 - \bar{\mathcal{M}}) \qquad [3.52]$$

where ψ denotes an eigenvalue of $\bar{\mathcal{M}}$.

Using these relationships, the spatio-temporal distribution of the envelope $E_0(z,t)$ and the reflection and transmission coefficients can be calculated. First of all, we seek the step response corresponding to an incident "wave", using the Heaviside step function $\Theta(t)$ in temporal terms. In $z = 0$, the Heaviside step is applied $\tilde{\mathcal{F}}(0,t) = \Theta(t)$, and in $z = L$ (where L is the thickness), there are no incident waves such that $\tilde{\mathcal{R}}(L,t) = 0$, which leads to:

$$\bar{\mathcal{E}}(L,t) \equiv \begin{pmatrix} \tilde{\mathcal{F}}(L,t) \\ 0 \end{pmatrix} = \bar{R}(L,t)\begin{pmatrix} \Theta(0^+) \\ \tilde{\mathcal{R}}(0,0^+) \end{pmatrix} \qquad [3.53]$$

It follows that:

$$\tilde{\mathcal{R}}(0,0^+) = -\frac{\overline{R_{21}}(L,t)}{\overline{R_{22}}(L,t)} \qquad [3.54]$$

In addition:

$$\bar{\mathcal{E}}(0,t) = \bar{R}(0,t)\begin{pmatrix} \Theta(0^+) \\ \tilde{\mathcal{R}}(0,0^+) \end{pmatrix} \qquad [3.55]$$

By combining these equations, we deduce:

$$\tilde{\mathcal{R}}(0,t) = \overline{R_{21}}(0,t) - \overline{R_{22}}(0,t)\frac{\overline{R_{21}}(L,t)}{\overline{R_{22}}(L,t)} \qquad [3.56]$$

Using the definition of the reflection coefficient, we deduce the step response in terms of reflectance:

$$R_{\Theta}(t) = \left|\tilde{\mathcal{R}}(0,t)\right]_0^t\Big|^2 \qquad [3.57]$$

In a similar way, the step response in terms of transmittance is given by:

$$R_{\Theta}(t) = \left|\tilde{\mathcal{F}}(L,t)]_0^t\right|^2 \quad [3.58]$$

with:

$$\tilde{\mathcal{F}}(L,t) = \frac{Det[\bar{R}(L,t)]}{\overline{R_{22}}(L,t)} \quad [3.59]$$

By analyzing the step response in terms of reflectance, it can be shown that it follows a "universal" curve presented in Figure 3.3 using the reduced time $\bar{t}$ given by:

$$\bar{t} = t\,\frac{\pi}{\Lambda_{ext}}\left(\frac{k}{c\,k_{\perp}}\right)^{1} \quad [3.60]$$

with the usual notation and Λ_{ext} denoting the extinction length of the 1D-PC.

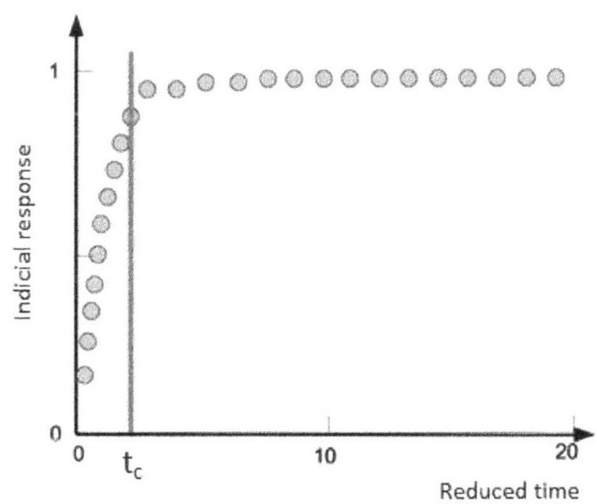

Figure 3.3. *Step response normalized to unity for the reflectance as a function of the reduced time*

We observe a transitory period up to the characteristic transition time t_c from which the system responds as described by stationary theory; in other words, the reflectance is given by the theory independent of the time. The characteristic time t_c is approximately equal to 2 units of reduced time.

In contrast, the step response in terms of transmission depends not only on the extinction length but also the absorption coefficients of the medium. Using the step response, the time response of the reflection and transmission of a pulse lasting for a "short" limited time can be calculated. Since the characteristic transition time t_c is typically of the order of a femtosecond, "short" means pulses that last in the order of a femtosecond.

If $S(t)$ denotes the time response of the reflected or transmitted pulse, the Laplace transform $S(s)$ of $S(t)$ is connected to the Laplace transform $E_0(s)$ of $E_0(t)$ by the convolution theorem:

$$S(s) = \hat{Z}_\delta(s)\, E_0(s) \qquad [3.61]$$

where $\hat{Z}_\delta(s)$ is the transfer function in terms of reflection or transmission, the Laplace transform of the pulse response $\hat{Z}_\delta(t)$ corresponding to a Dirac $\delta(t)$-type incident wave. $\hat{Z}_\delta(t)$ must not be confused with the step response $\hat{Z}_\Theta(t)$ where the incident wave is the Heaviside step $\Theta(t)$. Nevertheless, $\hat{Z}_\delta(t)$ can be determined from $\hat{Z}_\Theta(t)$ using various methods; one basic but effective method is Strejc's graphic method. We present an illustration of the transitory phenomenon in the hard X-ray domain (see Figure 3.4). The structure consists of a Bragg mirror made up of 100 Fe/C bilayers with a period d of 5.0 nm and a division ratio γ of 0.5. The incident radiation has an energy of 8 keV, which corresponds to a Bragg angle of approximately 0.93°. The stationary peak reflectivity is 64%. The incident pulse is a sine square $E_{0inc}(t) \sim sin^2\left(\frac{\pi t}{T}\right)$ in the time interval (0, T); in the example, T = 10 fs. We note that the reflected pulse only reaches approximately 8% of the stationary value.

Using the coupled-wave theory, we can also evaluate the reduction factor S_{vg} of the group velocity $\tilde{v}_g = \frac{d\omega}{dK}$ (which generalizes the group velocity v_g to the Bloch wave) with regard to the speed of light c:

$$S_{vg} \equiv \frac{c}{\tilde{v}_g} = \frac{1}{d} \frac{\int_0^d dz \left(|\tilde{\mathcal{F}}(z,t)|^2 - |\tilde{\mathcal{R}}(z,t)|^2\right)}{\left(|\tilde{\mathcal{F}}(z,t)|^2 + |\tilde{\mathcal{R}}(z,t)|^2\right)} \qquad [3.62]$$

For a sinusoidal optical potential, the factor S_{vg} is given by:

$$S_{vg} = I_0(2\sigma) \qquad [3.63]$$

where I_0 is the modified Bessel function of order 0. For a square potential, this factor is written as:

$$S_{vg} = \frac{\sinh(\pi\,\sigma)}{\pi\,\sigma} \qquad [3.64]$$

with $\sigma = \Delta n \frac{d}{2\,\lambda_B}$ at the Bragg wavelength λ_B. Δn is the real index contrast.

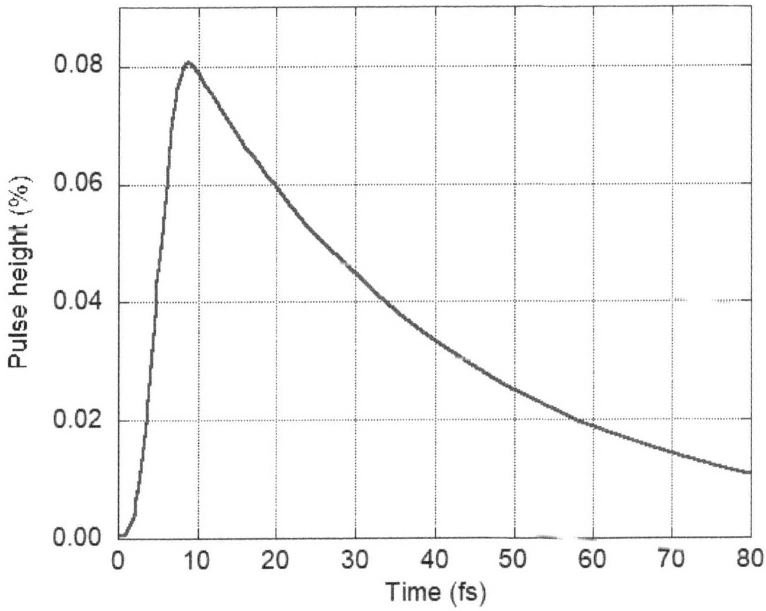

Figure 3.4. *Time evolution of pulse height for a sine square incident pulse of width 10 fs for the system being considered (see text)*

3.4. Generalization to Laue geometry

Although this book is about artificial Bragg optics, which implies a diffraction geometry in reflection mode, it is useful to also consider the so-called Laue transmission mode. In fact, the distinction is more about the culture in the field of spectroscopy of X-ray radiation but fundamentally the phenomena at play are the same: a periodic structure, with a period similar to the wavelength of the diffracted radiation, is implemented and the diffraction occurs by reflection on layers that are perpendicular or at a low angle to the external faces of the diffracting structure, and not parallel to them. Given the importance of this Laue mode in focusing and scattering applications, we dedicate this section to the application of coupled-wave theory to this mode of diffraction.

Recent developments of Laue optics for X-ray radiation (lenses and gratings) require an assessment of their diffraction efficiency performance, in particular so that they may be optimized. A series of works on this theme was published by Russian scientists (Levashov and Vinogradov 1993, 1994; Fechtchenko et al. 2002). Their approach is generally based on the so-called *two-wave approximation* (TWA) and provides very effective tools for optimization of the Laue optics concerned. Schneider presented a coupled-wave theory which goes beyond the TWA and is valid for absorbing media but restricted to the so-called first derivative approximation (Schneider 1997). This approach is frequently used by developers of Laue optics intended for soft X-ray radiation (Hambach et al. 2001; Huang et al. 2012).

Here, we present a more rigorous theory insofar as it conserves the second and first derivatives, is not limited to the TWA and is valid for absorbing media. This approach extends that used by Moharam and Gaylor (1981). Laue artificial systems are based on a periodic multilayer structure, cut and enlightened in parallel to the trench, as illustrated in Figure 3.5 which represents the geometry of the problem.

We suppose a *TE* polarization where the electric field $E(\boldsymbol{r}), \boldsymbol{r} = (x, z)$ is along the **y** axis, perpendicular to the $(\boldsymbol{x}, \boldsymbol{z})$ plane. This electric field obeys the Helmholz equation which in this case is written as:

$$\frac{\partial^2 E(r)}{\partial x^2} + \frac{\partial^2 E(r)}{\partial z^2} + k^2\, \varepsilon(\boldsymbol{r})\, E(\boldsymbol{r}) = 0 \qquad [3.65]$$

The solution is sought in the form of a development in plane waves:

$$E(\boldsymbol{r}) = \sum_m \mathcal{E}_m(x)\ exp(i\,\boldsymbol{\rho_m}.\boldsymbol{r})\,, \boldsymbol{\rho_m} = \boldsymbol{\rho_O} + m\,\boldsymbol{g}\ ;\ \boldsymbol{g} = \frac{2\pi}{d}\widehat{\boldsymbol{u}} \qquad [3.66]$$

The dielectric constant which is periodic along the $\widehat{\boldsymbol{u}}$ axis is expanded into a Fourier series:

$$\varepsilon(\boldsymbol{r}) = \sum_p \epsilon_p\ exp(i\,p\,\boldsymbol{g}.\boldsymbol{r}) \qquad [3.67]$$

Inserting equations [3.67] and [3.68] into equation [3.65], it becomes:

$$\sum_m [(\ddot{\mathcal{E}}_m(x) + 2\,i\,\rho_{mx}\,\dot{\mathcal{E}}_m(x) + (k^2\epsilon_0 - {\rho^2}_{mx})\,\mathcal{E}_m(x))\ exp(i\,m\,\boldsymbol{g}.\boldsymbol{r})] + k^2 \sum_m \sum_{p'} \epsilon_p\, \mathcal{E}_m(x) exp(i\,(p' + m)\boldsymbol{g}.\boldsymbol{r}) = 0\,;\, p' \neq 0 \qquad [3.68]$$

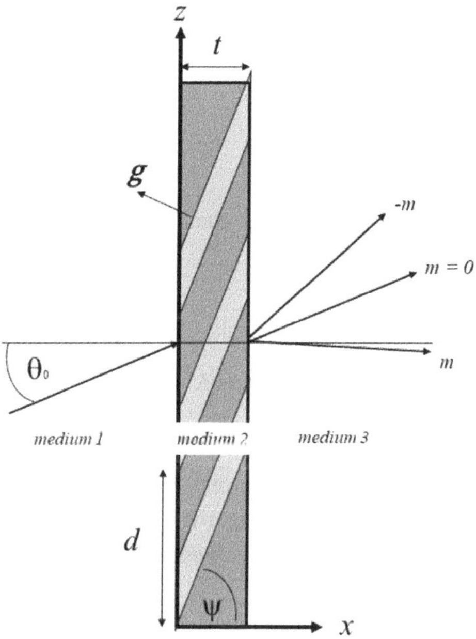

Figure 3.5. *The incident radiation comes from medium 1 by propagating in the $(\boldsymbol{x},\boldsymbol{z})$ plane and lights up the structure under an angle of incidence θ_0. For a color version of this figure, see www.iste.co.uk/andre/xrayradiation.zip*

COMMENT ON FIGURE 3.5.– *The radiation is diffracted within the grating volume (medium 2) with thickness t. The diffracted radiation (3 orders are represented) exits into medium 3. The unit vector $\hat{\mathbf{u}}$, which supports the reciprocal vector $\boldsymbol{g}$, is perpendicular to the stratification plane tilted by angle ψ with respect to the $\boldsymbol{x}$ axis.*

The dots $\dot{\cdot\cdot}$ denote derivation with respect to the variable x. The index x indicates the component on the $\boldsymbol{x}$ axis and p' is an integer not equal to zero $(p' \neq 0)$. By multiplying equation [3.68] by $exp(-\,i\,n\,\boldsymbol{g}.\boldsymbol{r})$ and integrating over a period, for the nth component, we obtain:

$$\ddot{\mathcal{E}}_n(x) + 2\,i\,\rho_{nx}\,\dot{\mathcal{E}}_m(x) + \left(k^2\epsilon_0 - \rho^2{}_{nx}\right)\mathcal{E}_n(x) + k^2\sum_{p'}\epsilon_p\;\mathcal{E}_{n-p'}(x) = 0 \qquad [3.69]$$

Relationship [3.69] comprises a system of coupled differential equations which is the basis for the coupled-wave theory for this configuration.

In fact, for a quasi-infinite structure along the $\boldsymbol{x}$ axis, $\mathcal{E}_m$ depends little on the variable x, such that $\ddot{\mathcal{E}}_m(x) \approx 0, \dot{\mathcal{E}}_m(x) \approx 0$, and we can write approximately:

$$\mathcal{E}_n = -k^2 \sum_{p'} \epsilon_p \frac{\mathcal{E}_{n-p'}}{k^2\epsilon_0 - \rho^2{}_{nx}} \qquad [3.70]$$

The term $\mathcal{E}_0$ is generally dominant, such that, in this case, we can write:

$$\mathcal{E}_n = k^2 \frac{\epsilon_n \, \mathcal{E}_0}{\rho^2{}_{nx} - k^2\epsilon_0} \qquad [3.71]$$

The strong coupling occurs for $\rho^2{}_{nx} - k^2\epsilon_0 \approx 0$ which corresponds to the Bragg condition.

A more rigorous calculation requires conservation of the derivatives $\ddot{\mathcal{E}}_m(x), \dot{\mathcal{E}}_m(x)$, which leads to a heavier mathematical development, an outline of which is given below. Resolving [3.69] involves a mathematical technique known as representation of state variables.

The state variables are introduced in the form of a column matrix:

$$\mathcal{E}_{\mathbf{0}}(x) \equiv \mathcal{E}(x), \mathcal{E}_{\mathbf{1}}(x) \equiv \dot{\mathcal{E}}(x), \mathcal{E}_{\mathbf{2}}(x) \equiv \ddot{\mathcal{E}}(x) = \dot{\mathcal{E}}_{\mathbf{1}}(x) \qquad [3.72]$$

Equation [3.69] is rewritten using these state variables and a “super-matrix” $\boldsymbol{S}$ as follows:

$$\dot{\mathcal{F}}(x) = \boldsymbol{S}\, \mathcal{F}(x) \qquad [3.73]$$

with:

$$\mathcal{F}(x) \equiv \begin{pmatrix} \mathcal{E}_{\mathbf{0}} \\ \mathcal{E}_{\mathbf{1}} \end{pmatrix}, \dot{\mathcal{F}}(x) = \begin{pmatrix} \mathcal{E}_{\mathbf{1}} \\ \mathcal{E}_{\mathbf{2}} \end{pmatrix}, \boldsymbol{S} = \begin{pmatrix} [0] & \mathbb{I} \\ \left[\frac{2\,\rho_{lx}\,\mathcal{M}_{lj}}{i}\right] & \frac{2\,\rho_{lx}\,\delta_{lj}}{i} \end{pmatrix} \qquad [3.74]$$

In the expression for the super-matrix $\boldsymbol{S}$ (whose dimension is in practice equal to the number of orders retained in the calculation), $\mathbb{I}$ denotes the identity matrix, $[a_{il}]$ the matrix constructed from the elements a_{ij}, δ_{lj} corresponds to the Kronecker symbol and $\mathcal{M}_{lj}$ are the elements of the matrix $\boldsymbol{\mathcal{M}}$ which connect $\mathcal{E}_{\mathbf{1}}$ to $\mathcal{E}_{\mathbf{0}}$, in other words:

$$\mathcal{E}_{\mathbf{1}} = \boldsymbol{\mathcal{M}}\, \mathcal{E}_{\mathbf{0}} \qquad [3.75]$$

The elements $\mathcal{M}_{lj}$ are obtained from equation [3.5] when we write $\mathcal{E}_2 = \mathbf{0}$; these elements are independent of the variable x. At this stage, the problem consists of resolving equation [3.72] with the appropriate limit conditions and the conditions of continuity of the tangential components of the electric field and the magnetic field. The limit conditions are given by:

– in medium 1 (see Figure 3.5): $E_{(1)}(\boldsymbol{r}) = exp\left[i\left(\zeta_{(1)0x}x + \zeta_{(1)0z}z\right)\right] + \sum_m \left(R_m\, exp\left[i\left(\zeta_{(1)mx}x + \zeta_{(1)mz}z\right)\right]\right)$;

– in medium 3 (see Figure 3.5): $E_{(3)}(\boldsymbol{r}) = \sum_m T_m\, exp\left[i\left(\zeta_{(3)mx}x + \zeta_{(3)mz}z\right)\right]$.

The quantities $\zeta_{(i)ms}$ are the components ($s = x$ or z) of the wave vector $\boldsymbol{\zeta}$ in the medium (i) for diffraction order m. The amplitudes R_m and T_m are the quantities to be determined at the end. The conditions of continuity are given by:

– continuity of the electric field in $x = 0$: $\mathcal{E}_\mathbf{0}(0) = [R_m + \delta_{0m}]$;

– continuity of the electric field in $x = t$: $\mathcal{E}_\mathbf{0}(t) = [T_m]$;

– continuity of the magnetic field in $x = 0$: $\mathcal{E}_\mathbf{1}(0) = \left[-i\,\zeta_{(1)mx}(R_m - \delta_{0m})\right]$;

– continuity of the magnetic field in $x = t$: $\mathcal{E}_\mathbf{1}(t) = \left[i\,\zeta_{(3)mx}T_m\right]$.

An effective calculation strategy is to write $\boldsymbol{\mathcal{F}}(x)$ in the form of an eigenvector expansion:

$$\boldsymbol{\mathcal{F}}(x) = \sum_k c_k\, \boldsymbol{W}_k\, e^{\lambda_k x} \tag{3.76}$$

where $\boldsymbol{W}_k$ is the kth eigenvector of $\boldsymbol{S}$ associated with the eigenvalue λ_k. The coefficients c_k must be obtained from the limit conditions of the problem. In practice, $\boldsymbol{\mathcal{F}}$ and $\boldsymbol{S}$ must be truncated. We will denote N the dimension of $\boldsymbol{\mathcal{F}}$. If we denote $\boldsymbol{M}$ the modal matrix formed from the N selected eigenvectors: $\boldsymbol{M} = \{\boldsymbol{W}_1, \dots, \boldsymbol{W}_N\}$ and $\boldsymbol{Z}$ the "shifted" modal matrix: $\boldsymbol{Z} = \left\{e^{\lambda_1 t}\boldsymbol{W}_1, \dots, e^{\lambda_N t}\boldsymbol{W}_N\right\}$, then:

$$\boldsymbol{\mathcal{F}}_t \equiv \boldsymbol{\mathcal{F}}(t) = \boldsymbol{Z}\,[c_k]\,;\ \boldsymbol{\mathcal{F}}_0 \equiv \boldsymbol{\mathcal{F}}(0) = \boldsymbol{M}\,[c_k] \tag{3.77}$$

which, by elimination of $[c_k]$, gives:

$$\boldsymbol{M}^{-1}\boldsymbol{\mathcal{F}}_0 - \boldsymbol{Z}^{-1}\boldsymbol{\mathcal{F}}_t = \boldsymbol{0} \qquad [3.78]$$

Since $\boldsymbol{\mathcal{F}}_0$ and $\boldsymbol{\mathcal{F}}_t$ are linear combinations of the amplitudes R_m and T_m, resolving [3.78] gives the efficiencies in terms of transmission of reflection.

If we consider that a strong coupling of the waves occurs primarily between order 0 and order m, TWA can be applied. Limiting ourselves to an approximation of the first derivative, it is then possible to deduce useful relationships to interpret the physical phenomena obtained using these Laue optics. In this case, the matrix $\boldsymbol{\mathcal{M}}$ is written using the book's usual notation:

$$\boldsymbol{\mathcal{M}} = i\begin{pmatrix} \frac{k^2\epsilon_0 - \rho_0{}^2}{2\,\rho_{0x}} & \frac{k^2\,\gamma\,\Delta\epsilon\, sinc(m\,\gamma\,\pi\,)}{2\,\rho_{0x}} \\ \frac{k^2\,\gamma\,\Delta\epsilon\, sinc(m\,\gamma\,\pi\,)}{2\,\rho_{mx}} & \frac{k^2\epsilon_0 - \rho_m{}^2}{2\,\rho_{mx}} \end{pmatrix} \qquad [3.79]$$

The solution to equation [3.75] is given by:

$$\mathcal{E}(x) = exp(\boldsymbol{\mathcal{M}}\,x)\,\mathcal{E}(0) \qquad [3.80]$$

The limit condition $\mathcal{E}(0) = \begin{pmatrix} 1 \\ \mathcal{E}_m(0) \end{pmatrix}$ associated with equation [3.80] leads to $\mathcal{E}_m(x)$:

$$\mathcal{E}_m(x) = i\frac{k^2}{2}\gamma\,\Delta\epsilon\, sinc(m\,\gamma\,\pi\,)\, exp(\chi\,z)\,\frac{sinh(\beta x)}{\beta\,\rho_{mx}} \qquad [3.81]$$

with:

$$\beta = i\,\frac{(2\,\mathrm{k}\,\gamma\Delta\epsilon\, sinc(m\,\gamma\,\pi\,))^{\,2}\rho_{0x}\rho_{mx} + (\rho_{0x}{}^2\rho_{mx} - \rho_{0x}\,\rho_{mx}{}^{\,2} + k^2\epsilon_0(\rho_{0x} - \rho_{mx})^2)}{4\,\rho_{0x}\rho_{mx}}$$

$$\chi = i\,\frac{k^2\epsilon_0(\rho_{0x} - \rho_{mx}) - (\rho_{0x}{}^2\rho_{mx} + \rho_{0x}\,\rho_{mx}{}^{\,2})}{\rho_{0x}\rho_{mx}} \qquad [3.82]$$

These last relationships extend those obtained by Levashov and Vinogradov (1994).

The diffraction efficiency in terms of transmission ε_{Tm} is given in the context of the previous approximations by $|\mathcal{E}_m(d)|^2$; in other words, explicitly:

$$\varepsilon_{Tm} = \frac{k^4}{4}\,|\gamma\,\Delta\epsilon\, sinc(m\,\gamma\,\pi\,)|^{\,2} exp(-\,Im(\chi)\,d)\,d^{\,2}\left|\frac{sinhc(\beta d)}{\rho_{mx}}\right|^2 \qquad [3.83]$$

The term $exp(-Im(\chi)\, d)$ gives the attenuation of the wave when traversing the structure; the term $|sinc(m\,\gamma\,\pi)|^2$ gives the extinctions similar to those of Bragg diffraction; the term $|sinhc(\beta d)|^2$ can be separated into two: the first $|sinc(Re(\beta d))|^2$ which expresses *Pendellösung*[1] oscillations; the second $|sinch(Im(\beta d))|^2$ which expresses the Borrmann–Campbell[2] anomalous transmission effect. Pendellösung oscillations in the Laue diffraction of multilayer trenches were mentioned in 2014 by André et al. (2014a) and discussed more recently by Punegov (2021).

Figure 3.6 provides a comparison of the results obtained by different theories. It presents the efficiency of diffraction in transmission as a function of the thickness t for the orders –1, 0, +1 and the following system: multilayer $MoSi_2$/Si with period $d = 100$ nm, division ratio $\gamma = 0.3$, $\psi = 0°$, radiation with wavelength $\lambda = 15$ nm, incident at angle $\theta_0 = 4.3°$. In the "rigorous" calculation (equations [3.73]–[3.78]), 15 modes (N = 15) are selected. We note that the curves differ significantly from one model to another. In the "rigorous" model, oscillations with a period equal to 7.5 nm appear (see Figure 3.6(c)), corresponding to $\frac{\lambda}{2}$, and due to interferences between the reflected waves on each face of the structure ($x = 0$ and $x = t$). These oscillations are not given by the other models considered. Generally, the Schneider model slightly overestimates the efficiency with respect to the "rigorous" calculation.

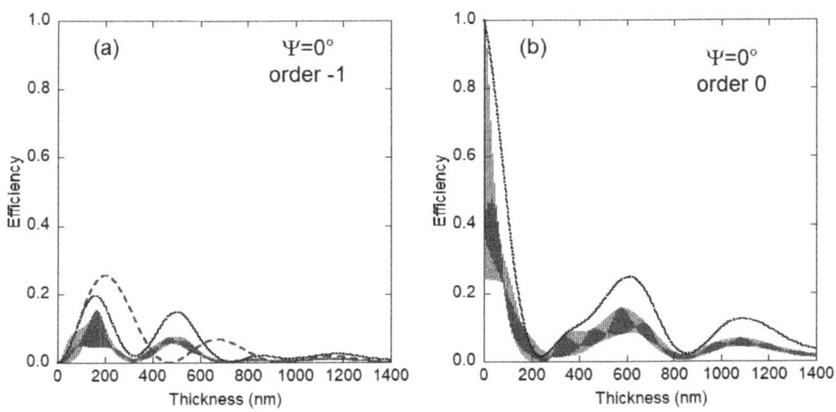

1. Pendellösung oscillations result from a beating phenomenon (interferences) due to phase modulation between the wave components in co-propagation in the structure (Ewald 1916).
2. The Borrmann–Campbell effect is an increase in intensity transmitted through a crystal when the Bragg condition is satisfied, and was theorized by von Laue (1949).

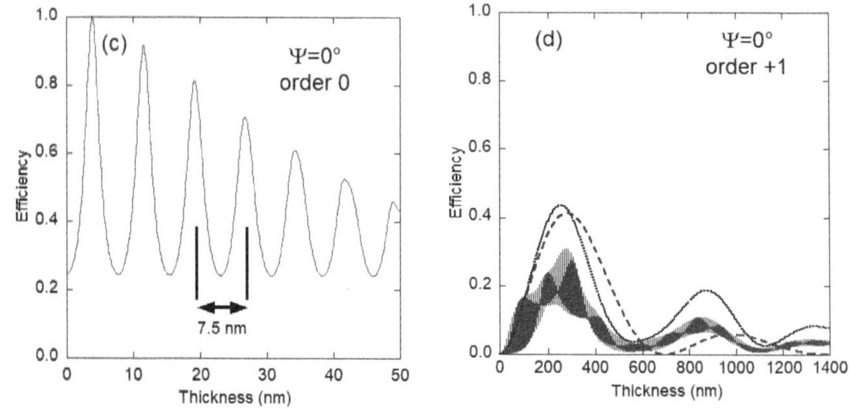

Figure 3.6. *Transmission efficiency as a function of thickness t of the structure whose characteristics are given in the text, at different orders. For a color version of this figure, see www.iste.co.uk/andre/xrayradiation.zip*

COMMENT ON FIGURE 3.6.– *(a) Order –1; (b) and (c) order 0; (d) order +1. "Rigorous" model red curve: "Schneider" model (Schneider 1997); blue curve and green curve: "Levashov–Vinogradov" model (Levashov and Vinogradov 1994)*[3].

3.5. References

André, J.-M. and Jonnard, P. (2015). Time-dependent Bragg diffraction and short-pulse reflection by one-dimensional photonic crystals. *Journal of Optics*, 17(8), 085609.

André, J.-M., Jonnard, P., Le Guen, K. (2013). Analysis by coupled-mode theory of X-ray multilayer gratings. In *UVX2012*, 02001. EDP Sciences.

André, J.-M., Le Guen, K., Jonnard, P. (2014a). Rigorous coupled-wave theory for lossy volume grating in Laue geometry X-ray spectroscopy [Online]. Available at: https://hal.science/hal-01082017 [Accessed 21 February 2023].

André, J.-M., Le Guen, K., Jonnard, P. (2014b). X-ray quasi-lamellar etched multilayers: Analysis by coupled-mode theory. *X-Ray Spectrometry*, 43(2), 122–125.

3. See, respectively, doi: 10.1063/1.120069 and doi: 10.1364/AO.32.001130.

Bourassin-Bouchet, C., Diveki, Z., de Rossi, S., English, E., Meltchakov, E., Gobert, O., Guénot, D., Carré, B., Delmotte, F., Ruchon, T. et al. (2011). Optiques pour les impulsions attosecondes. In *UVX 2010 – 10ème Colloque sur les Sources Cohérentes et Incohérentes UV, VUV et X; Applications et Développements Récents*. EDP Sciences, Ile de Porquerolles, 35–40 [Online]. Available at: http://uvx.edpsciences.org/10.1051/uvx/2011005 [Accessed 20 January 2023].

Chukhovskii, F.N. and Förster, E. (1995). Time-dependent X-ray Bragg diffraction. *Acta Crystallographica Section A: Foundations of Crystallography*, 51(5), 668–672.

Ewald, P.P. (1916). Zur Begründung der Kristalloptik. *Annalen der Physik*, 354(2), 117–143.

Fechtchenko, R.M., Vinogradov, A.V., Voronov, D.L. (2002). Optical properties of sliced multilayer gratings. *Optics Communications*, 210(3), 179–186.

Hambach, D., Schneider, G., Gullikson, E.M. (2001). Efficient high-order diffraction of extreme-ultraviolet light and soft X-rays by nanostructured volume gratings. *Optics Letters*, 26(15), 1200–1202.

Huang, Q., Li, H., Zhu, J., Wang, X., Wang, Z., Chen, L., Tang, Y. (2012). Fabrication and characterization of sliced multilayer transmission grating for X-ray region. *Chinese Optics Letters*, 10(9), 090501.

Kogelnik, H. (1969). Coupled wave theory for thick hologram gratings. *Bell System Technical Journal*, 48(9), 2909–2947.

Kolosov, S.I. and Punegov, V.I. (2005). Numerical integration methods for the Takagi-Taupin equations for crystals of rectangular cross section. *Crystallography Reports*, 50(3), 357–362.

Kozhevnikov, I.V., van der Meer, R., Bastiaens, H.M.J., Boller, K.-J., Bijkerk, F. (2011). Analytic theory of soft X-ray diffraction by lamellar multilayer gratings. *Optics Express*, 19(10), 9172–9184.

von Laue, M. (1949). Die Absorption der Röntgenstrahlen in Kristallen im Interferenzfall. *Acta Crystallographica*, 2(2), 106–113.

Levashov, V.E. and Vinogradov, A.V. (1993). Resonance diffraction efficiency enhancement in sliced multilayers. *Applied Optics*, 32(7), 1130–1135.

Levashov, V.E. and Vinogradov, A.V. (1994). Analytical theory of zone plate efficiency. *Physical Review E*, 49(6), 5797–5803.

Moharam, M.G. and Gaylord, T.K. (1981). Rigorous coupled-wave analysis of planar-grating diffraction. *JOSA*, 71(7), 811–818.

Morozov, G.V., Sprung, D.W.L., Martorell, J. (2004a). Semiclassical coupled wave theory for TM waves in one-dimensional photonic crystals. *Physical Review E*, 70(1), 016606.

Morozov, G.V., Sprung, D.W.L., Martorell, J. (2004b). Semiclassical coupled-wave theory and its application to TE waves in one-dimensional photonic crystals. *Physical Review E*, 69(1), 016612.

Osterhoff, M., Morawe, C., Ferrero, C., Guigay, J.-P. (2012). Wave-optical theory of nanofocusing x-ray multilayer mirrors. *Optics Letters*, 37(17), 3705–3707.

Punegov, V.I. (2021). X-ray Laue diffraction by sectioned multilayers. I. Pendellösung effect and rocking curves. *Journal of Synchrotron Radiation*, 28(5), 1466–1475.

Schneider, G. (1997). Zone plates with high efficiency in high orders of diffraction described by dynamical theory. *Applied Physics Letters*, 71(16), 2242–2244.

Taupin, D. (1964). Théorie dynamique de la diffraction des rayons X par les cristaux déformés. *Bulletin de Minéralogie*, 87(4), 469–511.

Taupin, D. (1967). Prévision de quelques images de dislocations par transmission des rayons X (cas de Laue symétrique). *Acta Crystallographica*, 23(1), 25–35.

Yan, H., Maser, J., Macrander, A., Shen, Q., Vogt, S., Stephenson, G.B., Kang, H.C. (2007). Takagi-Taupin description of X-ray dynamical diffraction from diffractive optics with large numerical aperture. *Physical Review B*, 76(11), 115438.

Yang, X., Kozhevnikov, I.V., Huang, Q., Wang, Z. (2015). Unified analytical theory of single-order soft x-ray multilayer gratings. *Journal of the Optical Society of America B*, 32(4), 506.

Yang, X., Kozhevnikov, I.V., Huang, Q., Wang, H., Hand, M., Sawhney, K., Wang, Z. (2017). Analytic theory of alternate multilayer gratings operating in single-order regime. *Optics Express*, 25(14), 15987–16001.

4

Propagation in Hybrid Bragg Structures: Scattering Theory

Jean-Michel ANDRÉ and Philippe JONNARD
LCPMR, CNRS, Sorbonne Université, Paris, France

4.1. Introduction

A short time after the techniques for deposition of layers at nanometric scale were mastered, allowing periodic multilayer mirrors operating as effective Bragg reflectors to be fabricated in the X-ray radiation domain, it became apparent that it would be useful to combine these techniques with other technologies in order to create other optical devices offering different or improved performances. The technologies implemented are taken primarily from those used in microelectronics (Khan and Malek 1991).

These devices combine the reflecting properties of Bragg mirrors with Fraunhofer or Fresnel diffractive properties. In 1963, Denisyuk (1963) proposed combining Bragg volume diffraction and Fraunhofer or Fresnel diffraction for the reconstruction of holographic images. Fraunhofer structures are diffraction gratings – bulk gratings covered with a multilayer Bragg structure (Barbee 1989; Puik et al. 1991), lamellar gratings (André et al. 1992) and alternate, echelette-type gratings (Rife et al. 1989, 1990) – constituting the so-called Bragg–Fraunhofer optics, whereas Fresnel optics are so-called Bragg–Fresnel lenses based on the concept of Fresnel zones, performing focusing or imaging functions (Aristov et al. 1988; Erko 1990). Periodic standard mirrors have also been divided into different forms:

periodic horizontal gradient mirrors (Nagel et al. 1982; Kortright et al. 1995; Stemmler et al. 1999), mirrors with horizontal gradient of ratio γ (Guichet et al. 1994), aperiodic mirrors, in particular super-mirrors (Morawe et al. 2002; Troussel et al. 2011) and "chirped" mirrors (Morlens et al. 2005; Diveki et al. 2014). These devices will be called hybrid Bragg structures.

To model the properties and performances of these hybrid structures, it is necessary to develop effective methods primarily from a numerical point of view. Indeed, using the Maxwell equations adapted to the specific geometry of optics, it is generally possible to find rigorous equations and to solve them using appropriate numerical methods and the initial and limit conditions. However, this generally results in considerable calculation times, associated with numerical instability problems, which is sufficient justification for seeking other approaches. For example, the calculation of the performances (efficiency of diffraction in reflection or transmission) of the gratings has been the subject of many studies in the context of a variety of theories (differential, integral, modal, etc.) (Maystre 1980; Petit 1980; Vincent 1980; Botten et al. 1981a, 1981b; Li 1993), which mainly present difficulties from a numerical point of view. Therefore, the differential (Nevière 1991) and modal (Sammar et al. 1991; Krastev et al. 1996) theories have been exploited in the domain of multilayer optics for X-ray radiation.

In this chapter, we present an approach inspired by the methods used in quantum mechanics and which have shown themselves to be very effective from a computational point of view, even though initially they involve relatively significant theoretical development. This approach presented in sections 4.2 and 4.3 is mainly taken and adapted from a series of works written by Sammar and André (1993a, 1993b, 1998).

4.2. Considerations on hybrid structures

The two main categories of hybrid structures are diffraction gratings and Fresnel lenses. In the context of the scattering theory developed in section 4.3, each structure is characterized by an optical potential, denoted as $P(\boldsymbol{R})$, connected to the spatial distribution of the dielectric constant $\varepsilon(\boldsymbol{R})$ of the structure considered in the relationship:

$$P(\boldsymbol{R}) = k^2(1 - \varepsilon(\boldsymbol{R})) \qquad [4.1]$$

4.2.1. *Grating structures*

This type of structure can be represented in quite a general way by the drawing in Figure 4.1.

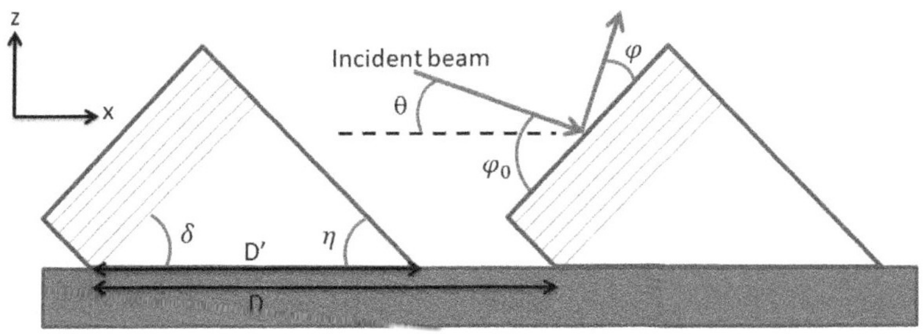

Figure 4.1. *Geometry of a multilayer grating*

For the mathematical notions and the notations, the reader can refer to Bracewell (2000). The potential in the reference framework presented in Figure 4.2 is given by:

$$P_g(\boldsymbol{R}) = P_g(x, z) = P_D(x, z) \otimes \sqcap_D(x) \quad [4.2]$$

The symbol $\otimes$ corresponds to the product of convolution and the symbol $\sqcap_X$ is the Dirac comb with period *X*. $P_D(x,z)$ is the optical potential of the horizontal unit cell given for a multilayer structure formed from a periodic arrangement of bilayers by:

$$P_D(x,z) = \sum_{n=0}^{N}(P_{\boldsymbol{a}} + P_{\boldsymbol{b}}\mathbf{T})\, \Theta\big(z - f_n(x)\big)\, \Theta\big(z - f(x)\big)\, \Theta(g_n(x) - z)\Theta(g(x) - z) \quad [4.3]$$

with:

$$f_n(x) = tan\delta\, x + n\, d\, ;\ g_n(x) = tan\delta\, x + (n\, d + d_a)\, f(x) = -tan\eta\, x\, ;\ g(x) = -tan\eta\, (x - D')P_a = k^2(1 - \varepsilon_a)\, ;\ P_b = k^2(1 - \varepsilon_b) \quad [4.4]$$

In equation [4.4], Θ is the Heaviside step function and **T** is a translation operator corresponding to the components vector $\{d_{\boldsymbol{a}}\, cos\delta, d_{\boldsymbol{a}}\, sin\delta\,\}$ in the reference frame ($\boldsymbol{x}$, $\boldsymbol{z}$). If $\delta = 0$ and $\eta = \frac{\pi}{2}$, it is a lamellar grating with width *D'*, which becomes a conventional mirror if $D = D'$. If $\delta \neq 0$, it is an echelette grating (sawtooth profile), sometimes called a "blazed" grating if the blaze condition is verified.

4.2.2. *Fresnel lens structures*

Fresnel lenses are based on a concept owed to the physicist Augustin Fresnel (Fresnel et al. 1866). We summarize the principle of these lenses by referring to Figure 4.2.

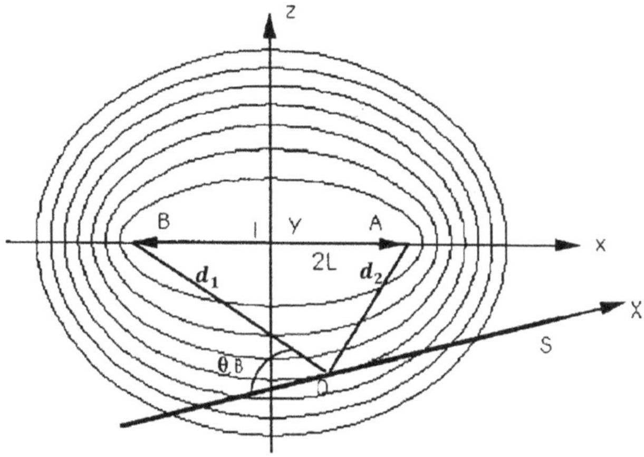

Figure 4.2. *Geometry of Fresnel lenses; the ellipses represent a projection of the equiphase ellipsoids in the plane of symmetry **(x,z)**. The distance from B to a point O is denoted as d_1 and from O to A is denoted as d_2 (according to André and Barchewitz 1994)*[1]

A monochromatic light source with wavelength λ is located at point B, and a point A is located on the **x** axis at a distance 2L from B. All the points O, such as the optical pathway BOA corresponding to a given phase, form an ellipsoid of revolution, or spheroid, with centers A and B, known as an equiphase spheroid. All the points P located between two equiphase spheroids associated with a path length difference equal to $\lambda/2$ form a geometric site **Λ** for which the phase term associated with the optical pathway BLA has the same sign. Let us state that A is the image of B using an optic **S** located on the **X** axis. In this case, we come to consider the intersections of the geometric sites **Λ** with the surface **S**. These intersections define the so-called Fresnel zones. It is possible to obtain an image of B in A by eliminating the contribution (amplitude regime) or inverting the phase (phase regime) of every other zone; because in this case, constructive interferences occur owing to the identical

1. See doi: 10.1051/anphys/1994003.

sign of the phase terms, without involvement of destructive interferences. For a plane surface **S**, the limits of the Fresnel zone have the following equations in the reference frame (I, **X**, **y**):

$$\frac{X^2}{K^2} + \frac{y^2}{K^2(T^2 - L^2)} = 1 \qquad [4.5]$$

where K and L are homogeneous quantities with lengths that are functions of the wavelength λ and geometric parameters (L, glancing angle θ of the rays on the surface **S**, etc.). Equation [4.5] defines a grating of *non-confocal* ellipses with a ratio of axes close to the sinus of angle θ. These ellipses degenerate into a circle in the case of a quasi-normal incidence or operation by transmission. In the case of the usual oblique incidence in Bragg–Fresnel optics, relationship [4.5] can be rewritten using the quantity f that corresponds to a focal distance and a quantity G corresponding to a magnification:

$$\left(X + m\,\frac{1-G}{1+G}\,\frac{cotg(\theta)}{sin(\theta)}\right)^2 + \frac{y^2}{sin^2(\theta)} = m\,\frac{2f\lambda}{sin^2(\theta)}\ ; m \in \mathbb{N}$$

$$f = \frac{d_1 d_2}{d_1 + d_2} \text{ and } G = \frac{d_2}{d_1} \qquad [4.6]$$

If we reduce our considerations to one-dimensional focusing, a point source with an image that is a right segment, the edges of the zones become right segments with distance D_m to the central axis of the lens given by:

$$D_m = \frac{\lambda}{m}\,\frac{cotg(\theta)}{sin(\theta)}\left(\sqrt{1 + tg^2(\theta)\left(1 + \frac{4f}{m\lambda}\right)} \pm 1\right) \qquad [4.7]$$

It is important to note the asymmetry of the distance of the lines from the central axis.

In the case of Bragg lenses, the incidence will be in the vicinity of the Bragg angle of the eponymous structure. The amplitude lenses are made by etching to the substrate of a Bragg structure in order to physically eliminate every other Fresnel zone.

In the following, we consider the case of the Fresnel zones with linear amplitude that result in lighter mathematical developments than the circular or elliptical zones. The drawing for such linear lenses is presented in Figure 4.3.

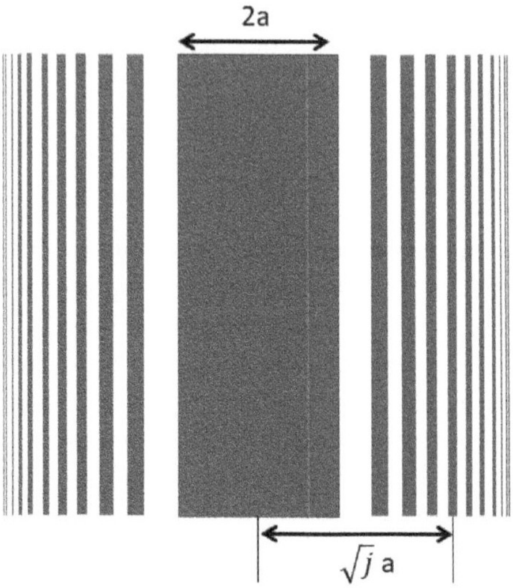

Figure 4.3. *Geometry of a linear Fresnel lens*

We can differentiate between positive lenses P and negative lenses N. In the positives, the blue zones in Figure 4.3 correspond to Bragg reflective zones and the white zones to a vacuum, and in the negatives the inverse is true. The optical potential will be constructed using the following profile functions with the notations shown in Figure 4.2:

- $\mathcal{F}_+(x) = rect\left[\frac{x}{a\sqrt{2j+1}}\right] - rect\left[\frac{x}{a\sqrt{2j}}\right]$ for P lenses;

- $\mathcal{F}_-(x) = rect\left[\frac{x}{a\sqrt{2(j+1)}}\right] - rect\left[\frac{x}{a\sqrt{2j+1}}\right]$ for N lenses.

The quantity j is a positive integer in the interval $[0, L]$, where L is the number of lines in the lens. The function *rect* is defined by:

$$rect\left[\frac{x}{b}\right] = \begin{cases} 1 \; if \; |x| \ll 1 \\ 0 \; if \; |x| > 1 \end{cases} \qquad [4.8]$$

We should mention that this geometry *is not optimal* as pointed out in works specializing in this type of structure (Aristov et al. 1988; Erko 1990) and as indicated in relationship [4.7]. Nevertheless, it provides an interesting basis for modeling these structures.

The optical potential $P_{F\pm}(x,z)$ of the P or N Fresnel lenses is written using $\mathcal{F}_{\pm}(x)$ as follows:

$$P_{F\pm}(x,z) = \mathcal{F}_{\pm}(x)\, Z(z) \qquad [4.9]$$

In the usual case where the lines are made up of a periodic multilayer structure formed from N bilayers **a** and **b**:

$$Z(z) = \sum_{n=0}^{N}(P_a + P_b\,\mathbf{T})\,\Theta(z - f_n)\Theta(g_n - z) \qquad [4.10]$$

where **T** is the translation operator corresponding to the coordinate vector $(0, d_a)$ and with the usual notations:

$$P_a = (1-\varepsilon_a)\,k^2, P_b = (1-\varepsilon_b)\,k^2, f_n = n\,d, g_n = d_a + nd \qquad [4.11]$$

4.3. Scattering theories applied to hybrid structures

4.3.1. *General approach*

The approach consists of considering a scattering optical environment characterized by an optical potential $P(\boldsymbol{R})$ (connected to the dielectric constant $\varepsilon(\boldsymbol{R})$, see section 4.2), in order to rewrite the Helmholtz equation in a Hamiltonian form usual in quantum mechanics using $P(\boldsymbol{R})$ and to use mathematical tools developed and used in quantum mechanics to resolve the scattering problems. Figure 4.4 shows this approach.

The electromagnetic field is represented by the scalar field $\psi(\boldsymbol{r})$, which obeys the Helmholtz equation:

$$\big(\Delta + k^2\,\varepsilon(\boldsymbol{R})\big)\psi(\boldsymbol{r}) = 0 \qquad [4.12]$$

This can be rewritten in the usual form for quantum mechanics:

$$(E - H_0)\psi(\boldsymbol{r}) = P(\boldsymbol{R})\psi(\boldsymbol{r})\,;\; E = k^2, H_0 = -\Delta \qquad [4.13]$$

The point $\boldsymbol{R}$ moves through the scattering spatial region **D**. If we note that in the X-ray radiation domain, $\varepsilon(\boldsymbol{R}) \cong 1$, the quantity $P(\boldsymbol{R})$ is small compared to E and can be considered to be a disturbance term.

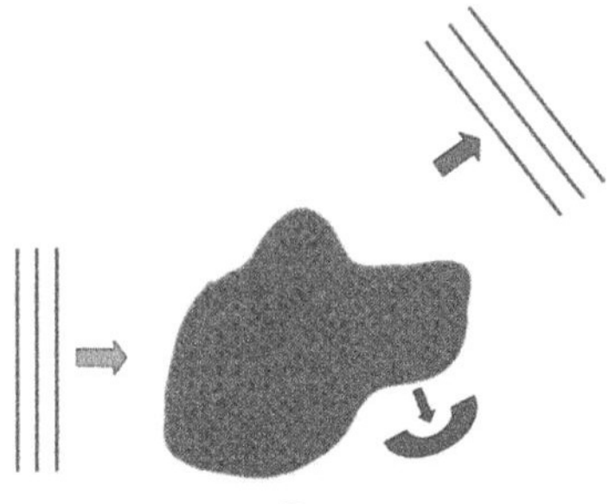

Figure 4.4. *A plane incident wave (green) interacts with the scattering optical medium* **D***; there are resulting scattered waves that are detected in the "near" (Fresnel, magenta color) or far-off (Fraunhofer, red color) field. For a color version of this figure, see www.iste.co.uk/andre/xrayradiation.zip*

Equation [4.13] is an inhomogeneous differential equation that can be formally solved using the Green method (Economou 1983) to give a Lippmann–Schwinger equation:

$$\psi^{\pm} = \psi_0 + G_0^{\pm}(E)\, P\, \psi^{\pm} \qquad [4.14]$$

where ψ_0 is the solution to the homogeneous equation $(E - H_0)\psi_0 = 0$ and $G_0^{\pm}(E)$ is the Green operator associated with the Green function $G_0^{\pm}(\boldsymbol{R}, \boldsymbol{R}', E)$ defined by:

$$(E - H_0)G_0^{\pm}(\boldsymbol{R}, \boldsymbol{R}', E) = \delta(\boldsymbol{R} - \boldsymbol{R}') \qquad [4.15]$$

The Green function $G_0^{\pm}(\boldsymbol{R}, \boldsymbol{R}', E)$ is given by (Barton 1989):

$$G_0^{\pm}(\boldsymbol{R}, \boldsymbol{R}', E) = -\frac{exp(\pm\, i\, k\, |\boldsymbol{R}-\boldsymbol{R}'|)}{4\,\pi|\boldsymbol{R}-\boldsymbol{R}'|} \qquad [4.16]$$

The Green function that is physically pertinent is $G_0^{+}(\boldsymbol{R}, \boldsymbol{R}', E)$. When iterating equation [4.11], it is useful to introduce the operator $T^{+}(E) = P + P\, G_0^{+}(E)\, P + P\, G_0^{+}(E)\, P\; G_0^{+}(E)\, P + \cdots$ which verifies:

$$T^{+}(E) = P + P\; G_0^{+}(E)\, T^{+}(E) \qquad [4.17]$$

The physical solution $\psi^{+}(\boldsymbol{r})$ is then obtained by identifying ψ_0 with the incident wave $exp(i\,\boldsymbol{k}.\boldsymbol{r})$ and by combining equations [4.14], [4.16] and [4.17], which gives:

$$\psi^{+}(\boldsymbol{r}) = exp(i\,\boldsymbol{k}.\boldsymbol{r}) - \frac{1}{4\,\pi}\int \frac{exp(\,i\,k\,|\boldsymbol{R}-\boldsymbol{r}|)}{|\boldsymbol{R}-\boldsymbol{r}|}\, T^{+}(\boldsymbol{R}, E)\, exp(i\,\boldsymbol{k}.\boldsymbol{R})\; d\boldsymbol{R} \qquad [4.18]$$

At this stage, two cases must be considered: diffraction in the so-called Fraunhofer far field and diffraction in the Fresnel nearer field (see Figure 4.3). The first case is obtained by keeping the first term in the expansion of $\frac{exp(i\,k\,|\boldsymbol{R}-\boldsymbol{r}|)}{|\boldsymbol{R}-\boldsymbol{r}|}$ in R/r, whereas the second requires a second-order expansion. If we want to take the two diffraction cases overall, a second-order expansion in R/r therefore becomes essential, which for $\psi^+(\boldsymbol{r})$ gives:

$$\psi^+(\boldsymbol{r}) =$$
$$exp(i\,\boldsymbol{k}.\boldsymbol{r}) - \frac{1}{4\pi}\frac{exp(i\,k\,r)}{k\,r}\int exp(-\,i\,\boldsymbol{k_d}.\boldsymbol{R}\,)\,exp\left(\frac{i\,k'R^2}{2\,r}\right)\,T^+(\boldsymbol{R},E)\,exp(i\,\boldsymbol{k}.\boldsymbol{R})\;d\boldsymbol{R} \quad [4.19]$$

with:

$$\boldsymbol{k_d} = k\,\frac{\boldsymbol{r}}{r}, k' = \frac{|\boldsymbol{k_d}\times\boldsymbol{R}|^2}{k\,R^2} \quad [4.20]$$

Using Dirac's "bra-ket" notation, we can rewrite equation [4.12] using the closure relationship[2] in the form:

$$\psi^+(\boldsymbol{r}) = exp(i\,\boldsymbol{k}.\boldsymbol{r}) - \frac{1}{4\pi}\frac{exp(i\,k\,r)}{k\,r}\,\langle\boldsymbol{k_d}|\widetilde{T^+}(\,E)|\boldsymbol{k}\rangle \quad [4.21]$$

where:

$$\widetilde{T^+}(\,E) = \tilde{P} + \tilde{P}\,G_0^+(E)\,\tilde{P} + \tilde{P}\,G_0^+(E)\,\tilde{P}\,G_0^+(E)\,\tilde{P}\;...\;;\;\tilde{P}\; = exp\left(\frac{i\,k'R^2}{2\,r}\right)\,P \quad [4.22]$$

A relationship similar to equation [4.17] relative to $T^+(E)$ applies to $\widetilde{T^+}(E)$, that is:

$$\widetilde{T^+}(E)\; = \tilde{P} + \tilde{P}\;\,G_0^+(E)\,\widetilde{T^+}(E) \quad [4.23]$$

Physically, the pertinent quantity is the scattering amplitude $\tilde{f}(\boldsymbol{k_d},\boldsymbol{k},r)$ given by:

$$\tilde{f}(\boldsymbol{k_d},\boldsymbol{k},r) = \langle\boldsymbol{k_d}|\widetilde{T^+}(E)|\boldsymbol{k}\rangle \quad [4.24]$$

To deal with Fraunhofer diffraction, we use $f(\boldsymbol{k_d},\boldsymbol{k},r)$, which is deduced from $\tilde{f}(\boldsymbol{k_d},\boldsymbol{k},r)$ by replacing $\widetilde{T^+}(E)$ with $T^+(E)$. Generally, all relationships valid for

2. The closure relationship in the direct space stipulates that $\int d\boldsymbol{R}\;|\boldsymbol{R}><\boldsymbol{R}| = 1$; a similar relationship applies in the reciprocal space $\int d\boldsymbol{K}\;|\boldsymbol{K}><\boldsymbol{K}| = 1$.

Fresnel diffraction (denoted by the "tilde" symbol $\tilde{...}$) apply to Fraunhofer diffraction when the $\tilde{}$ symbol is omitted.

At this stage, it is relevant to work in the reciprocal space $\boldsymbol{K}$. Using relationships [4.23] and [4.24] and the closure relationship in the reciprocal space, we obtain:

$$\tilde{f}(\boldsymbol{k_d}, \boldsymbol{k}, r) = -\frac{1}{4\pi}\hat{\tilde{P}}(\boldsymbol{Q}, r) + \lim_{s \to 0^+} \frac{1}{(2\pi)^d} \int \frac{\hat{\tilde{P}}(\boldsymbol{Q}, r)}{K^2 - k^2 + i\,s} \tilde{f}(\boldsymbol{K}, \boldsymbol{k}, r) d\boldsymbol{K}$$

$$\boldsymbol{Q} = \boldsymbol{k_d} - \boldsymbol{k} \qquad [4.25]$$

To obtain relationship [4.25], the representation of the Green function in the reciprocal space is taken into account[3] (Barton 1989):

$$G_0^+ = \lim_{s \to 0^+} \frac{-1}{K^2 - k^2 + i\,s} \qquad [4.26]$$

The symbol $\hat{...}$ denotes the Fourier transform:

$$\hat{F}(\boldsymbol{K}) = \int F(\boldsymbol{R}) \, exp(i\, \boldsymbol{K}.\boldsymbol{R}) d\boldsymbol{R} \qquad [4.27]$$

From the above, it follows that at order 1 of the expansion, the scattering amplitude is written as:

$$f^{(1)}(\boldsymbol{k_d}, \boldsymbol{k}) = -\frac{1}{4\pi}\hat{\tilde{P}}(\boldsymbol{Q}) \qquad [4.28]$$

At order 2:

$$f^{(2)}(\boldsymbol{k_d}, \boldsymbol{k}, r) = f^{(1)}(\boldsymbol{k_d}, \boldsymbol{k}, r) - \lim_{s \to 0^+} \frac{1}{(2\pi)^d} \frac{1}{4\pi} \int \frac{\hat{P}(\boldsymbol{k_d} - \boldsymbol{K})\hat{\tilde{P}}(\boldsymbol{K} - \boldsymbol{k}, r)}{K^2 - k^2 + i\,s} d\boldsymbol{K} \qquad [4.29]$$

and by recurrence:

$$f^{(n)}(\boldsymbol{k_d}, \boldsymbol{k}, r) = f^{(n-1)}(\boldsymbol{k_d}, \boldsymbol{k}, r) - \lim_{s \to 0^+} \frac{1}{(2\pi)^d} \int \frac{\hat{P}(\boldsymbol{k_d} - \boldsymbol{K}) f^{(n-1)}(\boldsymbol{K}, \boldsymbol{k}, r)}{K^2 - k^2 + i\,s} d\boldsymbol{K} \qquad [4.30]$$

This last relationship, subject to convergence to be determined, should allow an improvement of the precision in the calculation of the scattering amplitude.

3. d denotes the space dimension in which the calculation is performed; in cases generally considered in this chapter, $d = 2$.

4.3.2. *Application to Bragg–Fraunhofer diffraction*

We can apply the theoretical elements of section 4.3.1 to different cases of Bragg–Fraunhofer diffraction. This situation corresponds to Figure 4.1. The efficiency is calculated at the first order of the expansion using relationship [4.28].

This requires the quantity $\hat{\hat{P}}(\boldsymbol{K})$ to be determined, which in this case is written as:

$$\hat{\hat{P}}(\boldsymbol{K}) = \int_{-\infty}^{+\infty} dx \int_{-\infty}^{+\infty} P_D(x,z)\, exp\big(i(K_x\, x + K_z\, z)\big) \sqcap_G(K_x)\; dz\,;\, G = \frac{2\,\pi}{D} \qquad [4.31]$$

Introducing the expression for $P_D(x,z)$ given by equation [4.3], after a calculation using the properties of the Fourier transform, we obtain.

$$\hat{\hat{P}}(\boldsymbol{K}) = \hat{\hat{P}}(K_x, K_z) = \sum_{n=-\infty}^{+\infty} \delta_k(K_x - n\,G)\, A(K_x, K_z) \qquad [4.32]$$

with δ_k, the Kronecker symbol, and:

$$\begin{aligned} &A(K_x, K_z) = \Xi_+[tan(\delta), tan(\gamma)\,\alpha\, D', \alpha, tan(\delta)]\; U_{N+1}\big[\vartheta[K_x, K_z]\big] \\ &exp\big[i\,N\,\vartheta[K_x, K_z]\big]\big\{\big(P_a\,\Xi_-[tan(\gamma), \alpha\, d_a{}^*, 1, tan(\gamma)] + \\ &\exp\big[-i\,\sigma[K_x, K_z]\big]\big)\, P_b\,\Xi_-[tan(\gamma), \alpha\, d_b{}^*, 1, tan(\gamma)]\big\} \end{aligned} \qquad [4.33]$$

where:

$$\Xi_\pm[A, B, C, D] = \frac{1 - exp(-i(K_x \pm K_z A)\, B)}{C\,(K_x \pm K_z D)}$$

$$U_M[x] = \frac{sin(M\,x)}{sin(x)}$$

$$\vartheta[K_x, K_z] = \big(K_x - K_z\, tan(\gamma)\big)\,\alpha\,\frac{d^*}{2}$$

$$\sigma[K_x, K_z] = \Big(\big(K_x sin(\delta) - K_z\, cos(\delta)\big)\,\alpha\, d_a{}^*\Big)$$

$$d_x{}^* = d_x / cos(\delta)$$

$$\alpha = \frac{1}{tan(\delta) + tan(\gamma)} \qquad [4.34]$$

Equations [4.32]–[4.34] are long because they express all the cases shown in Figure 4.1 globally. They allow the diffraction efficiency $\epsilon^{(1)}(p)$ to be calculated to the first order, at order p:

$$\epsilon^{(1)}(p) = \left[\frac{1}{4\,\pi}\right]^2 A(Q_x, Q_z) \qquad [4.35]$$

Taking into account the Kronecker symbol in equation [4.32] and the conservation of energy:

$$Q_x = p\,G$$

$$Q_z = \pm\left(k^2 - \left({k_z}^2 + p\,G\right)^2\right) - k_z \qquad [4.36]$$

where the + sign corresponds to transmitted waves and the – sign corresponds to reflected waves. Using the series of the preceding equations, the kinematic equations that govern the diffraction conditions of these gratings can be found.

Firstly, considering the Dirac comb in equation [4.31], we obtain $Q_x = k_{dx} - k_x = p\,G$, so in terms of angles:

$$D\left(cos(\theta) - cos(\theta_0)\right) = p\,\lambda \qquad [4.37]$$

This relationship is none other than the grating diffraction formula. The Bragg diffraction condition is obtained by seeking the maxima of the function U_M in equation [4.34], so:

$$\left(Q_x - Q_z\,tan(\gamma)\right)\alpha\,\frac{d^*}{2} = m\,\pi, m\,\in \mathbb{N} \qquad [4.38]$$

or in angular terms:

$$d\left(sin(\varphi) - sin(\varphi_0)\right) = m\,\lambda \qquad [4.39]$$

Specific cases:

– laminar grating: $\delta = 0, \theta = \varphi, \theta_0 = \varphi_0$. By combining in this case equations [4.37] and [4.39], we obtain the generalized Bragg law (Warburton 1990):

$$m\,\lambda = d\,sin(\theta_0)\left[1 + \sqrt{1 - \frac{cos(\theta_0)}{sin^2(\theta_0)}\frac{p\,\lambda}{m\,D} - \frac{1}{sin^2(\theta_0)}\left(\frac{p\,\lambda}{m\,D}\right)^2}\right] \qquad [4.40]$$

Making D tend towards ∞, we return to the standard Bragg law:

$$m\,\lambda = 2\,d\,sin(\theta_0) \qquad [4.41]$$

– echelette grating: $\delta \neq 0, \theta_0 \neq \varphi_0$. The maximum efficiency of this type of grating will be achieved at the so-called blaze condition obtained by canceling the argument D of the function Ξ_+ in equation [4.34], so:

$$Q_x + Q_z\,tan(\delta) = 0 \qquad [4.42]$$

which results in $\varphi = \varphi_0$, that is, the condition of specular reflection on the grating facets. By associating this condition with equation [4.39], we achieve the blaze condition in terms of the angle of incidence φ_0:

$$m\,\lambda = 2\,d\,sin({\varphi_0}^{bl}) \qquad [4.43]$$

and the grating law is then written in terms of the blaze angle of incidence ${\varphi_0}^{bl}$:

$$p\,\lambda = 2\,Dsin(\delta)sin({\varphi_0}^{bl}) \qquad [4.44]$$

To find agreement between the blaze condition, the grating law and the generalized Bragg law, the following condition must be met (Rife et al. 1989):

$$\frac{Dsin(\delta)}{d} = \frac{p}{m} \qquad [4.45]$$

Table 4.1 gives the efficiency of peak diffraction at various orders for a laminar grating with the characteristics given in the key. The results indicate the remarkable precision provided by the scattering theory for the expansion taken to the second order.

The resolvent method (Reuse 2007; Cohen-Tannoudji et al. 2012) can be used to deal with relationship [4.25] in relation to the case of Bragg–Fraunhofer diffraction. We write:

$$f(\boldsymbol{k_d}, \boldsymbol{k}) = -\frac{1}{4\,\pi}\,\hat{P}(\boldsymbol{Q}, r) - \mathbb{A}[\,f(\boldsymbol{K}, \boldsymbol{k})] \qquad [4.46]$$

where $\mathbb{A}$ is the integral operator defined by:

$$\mathbb{A}[\,f(\boldsymbol{K}, \boldsymbol{k})] = \frac{1}{(2\pi)^d}\int \frac{\hat{P}(\boldsymbol{K}, \boldsymbol{k_d})}{K^2 - k^2}\,f(\boldsymbol{K}, \boldsymbol{k})\,d\boldsymbol{K} \qquad [4.47]$$

Order	(a)	(b)	(c)	(d)	(e)
-2	0.0346	0	0.0380	–	9
-1	0.3937	0.4440	0.3935	11	0.05
0	1.0450	1.1790	1.0463	11	0.12
+1	0.4534	0.5143	0.4528	12	0.13
+2	0.0346	0	0.0533	–	35

Table 4.1. *Efficiencies of a diffraction grating*

COMMENT ON TABLE 4.1.– *(a) Absolute efficiency using a "rigorous" dynamic method; (b) absolute efficiency using the first-order scattering method; (c) absolute efficiency using the second-order scattering method; (d) relative error between the "rigorous" theory and the first-order scattering theory; (e) relative error between the "rigorous" theory and the second-order scattering theory; the data are in percentages. The case considered is a laminar grating with a Bragg structure with periodic stacking of bilayers: material **a** (molybdenum); $d_a = 1$ nm; material **b** (carbon); $d_b = 2$ nm and the parameters: $D = 360$ nm; $D'/D = G = 0.5$; radiation energy 930 eV.*

If the optical potential $P(\boldsymbol{R})$ can be divided into two terms:

$$P(\boldsymbol{R}) = P_0(z) + P_\lambda(x, z) \qquad [4.48]$$

then $\mathbb{A}$ will also be divided into two terms:

$$\mathbb{A} = \mathbb{A}_0 + \mathbb{A}_\lambda \qquad [4.49]$$

the first $\mathbb{A}_0$ acting on P_0 and the second $\mathbb{A}_\lambda$ acting on P_λ.

In addition, if the resolvent $\mathbb{R}$ is introduced from equation [4.46] by:

$$f(\boldsymbol{k}_d, \boldsymbol{k}) = -\frac{1}{4\pi}\,\mathbb{R}\big[\hat{P}(\boldsymbol{k}_d, \boldsymbol{k})\big] \qquad [4.50]$$

then it follows that:

$$\mathbb{R} = (\mathbb{I} + \mathbb{A})^{-1} = (\mathbb{I} + \mathbb{A}_0 + \mathbb{A}_\lambda)^{-1} \qquad [4.51]$$

with $\mathbb{I}$ the identity operator, so after some algebra:

$$\mathbb{R} = (\mathbb{I} + \mathbb{R}_0\, \mathbb{A}_\lambda)^{-1} \mathbb{R}_0 \cong \mathbb{R}_0 (\mathbb{I} - \mathbb{A}_\lambda \mathbb{R}_0) \qquad [4.52]$$

where we can state:

$$\mathbb{R}_0 = (\mathbb{I} + \mathbb{A}_0)^{-1} \qquad [4.53]$$

By applying the previous formalism, relationship [4.50] can be rewritten in the form:

$$f(\boldsymbol{k_d}, \boldsymbol{k}) = f_0(\boldsymbol{k_d}, \boldsymbol{k}) + f_\lambda(\boldsymbol{k_d}, \boldsymbol{k}) \qquad [4.54]$$

with:

$$f_0(\boldsymbol{k_d}, \boldsymbol{k}) = \frac{1}{4\pi} \mathbb{R}_0 \left[\widehat{P_0}(\boldsymbol{k_d}, \boldsymbol{k})\right]$$

$$f_\lambda(\boldsymbol{k_d}, \boldsymbol{k}) = \frac{1}{4\pi} \left(\mathbb{R}_0 \left[\widehat{P_\lambda}(\boldsymbol{k_d}, \boldsymbol{k})\right] - \mathbb{R}_0\, \mathbb{A}_\lambda \mathbb{R}_0 \left[\widehat{P_0}(\boldsymbol{k_d}, \boldsymbol{k})\right]\right) \qquad [4.55]$$

At this stage, the problem is now to determine $f_\lambda(\boldsymbol{k_d}, \boldsymbol{k})$. We rewrite $f_\lambda(\boldsymbol{k_d}, \boldsymbol{k})$ in extended form:

$$f_\lambda(\boldsymbol{k_d}, \boldsymbol{k}) = \mathbb{R}_0 \left(\frac{1}{4\pi}\left[\widehat{P_\lambda}(\boldsymbol{k_d}, \boldsymbol{k})\right] - \frac{1}{(2\pi)^d} \int \frac{\widehat{P_\lambda}(\boldsymbol{K}, \boldsymbol{k_d})}{k^2 - K^2} f_0(\boldsymbol{K}, \boldsymbol{k})\, d\boldsymbol{K}\right) \qquad [4.56]$$

Returning to the direct space in the integrand:

$$f_\lambda(\boldsymbol{k_d}, \boldsymbol{k}) = \frac{1}{4\pi} \mathbb{R}_0 \Big(\left[\widehat{P_\lambda}(\boldsymbol{k_d}, \boldsymbol{k})\right] - \int P_\lambda(\boldsymbol{R})\, exp(-i\, \boldsymbol{k_d}\, \boldsymbol{R})\, (G_0(\boldsymbol{R}) \otimes [P_0(\boldsymbol{R})\, \psi_0(\boldsymbol{R})])\, d\boldsymbol{R}\Big) \qquad [4.57]$$

In equation [4.57], $\otimes$ denotes the convolution product, $G_0(\boldsymbol{R})$ is the Green function in the direct space (see equation [4.16]) and ψ_0 is the field corresponding to the potential P_0, which can be written as:

$$\psi_0(\boldsymbol{R}) = exp(-i\, k_{//}\, \rho)\, \psi_z(z) \qquad [4.58]$$

where $\rho = (x, y)$ and $k_{//} = (k_x, k_y)$. The field $\psi_z(z)$ is the solution to the one-dimensional Helmholtz equation according to the variable z.

By applying the Lippmann–Schwinger equation to ψ_0, we obtain:

$$f_\lambda(\boldsymbol{k_d}, \boldsymbol{k}) = \frac{1}{4\pi} \mathbb{R}_0 \Big(\left[\widehat{P_\lambda}(\boldsymbol{k_d}, \boldsymbol{k})\right] - \int P_\lambda(\boldsymbol{R})\, exp(-i\, \boldsymbol{k_d}\, \boldsymbol{R})\, [exp(i\, \boldsymbol{k}.\boldsymbol{R}) - \psi_0(\boldsymbol{R})]\, d\boldsymbol{R}\Big) \qquad [4.59]$$

so:

$$f_\lambda(\boldsymbol{k_d}, \boldsymbol{k}) = \frac{1}{4\pi}\mathbb{R}_0\left(\int P_\lambda(\boldsymbol{R})\, exp(-i\, \boldsymbol{k_d}\, \boldsymbol{R})\, \psi_0(\boldsymbol{R})\, d\boldsymbol{R}\right) \qquad [4.60]$$

which is expressed in the reciprocal space as:

$$f_\lambda(\boldsymbol{k_d}, \boldsymbol{k}) = \frac{1}{4\pi}\mathbb{R}_0\left(\int \widehat{P_\lambda}(\boldsymbol{K} - \boldsymbol{k_d})\, \widehat{\psi_0}(\boldsymbol{K})\, d\boldsymbol{K}\right) \qquad [4.61]$$

Using the Fubini theorem:

$$f_\lambda(\boldsymbol{k_d}, \boldsymbol{k}) = \int \frac{\widehat{\psi_0}(\boldsymbol{K})}{4\pi}\mathbb{R}_0\widehat{P_\lambda}(\boldsymbol{K} - \boldsymbol{k_d})\, d\boldsymbol{K} \qquad [4.62]$$

With this approach, which can appear a little long from a mathematical point of view, the problem can be simplified considerably from a numerical point of view once condition [4.48] is satisfied. Now, this condition is generally met for Bragg–Fraunhofer gratings provided that conical diffraction is not occurring, in other words provided that the plane of incidence is perpendicular to the lines of the grating, which gives the previous approach a large amount of scope. Let us consider the optical potential in the latter case. It can be written as:

$$P(\boldsymbol{R}) = k^2(\varepsilon(z) - 1)V(x)\,; V(x) = \mu + M(x) \qquad [4.63]$$

$M(x)$ expresses modulation of the optical potential along the **x** axis, determined by the grating profile, and μ is the average value of this modulation obtained as the average value for a period:

$$\mu = \frac{1}{D}\int_{-D/2}^{+D/2} V(x)dx \qquad [4.64]$$

The quantity μ is none other than the constant term in the Fourier expansion of $V(x)$. Finally, we can write:

$$P(\boldsymbol{R}) = P_0(z)\left(1 + \frac{1}{\mu}\, M(x)\right)\,;\; P_0(z) = \mu\, k^2(\varepsilon(z) - 1) \qquad [4.65]$$

By combining equations [4.45], [4.59] and [4.62], we obtain:

$$f_\lambda(\boldsymbol{k_d}, \boldsymbol{k}) = \frac{\widehat{M}(k_x - k_{dx})}{\mu}\int \psi_{kz}(z)\, P_0(z)\, \psi_{-kdz}(z)\, dz \qquad [4.66]$$

where we have taken into account the symmetry condition $f(K_z, -k_{dz}) = f(k_{dz}, -K_z)$.

By virtue of equation [4.66], it becomes possible to calculate the scattering amplitude by simply integrating along the **z** axis after determining $\psi_{kz}(z)$ and $\psi_{-kdz}(z)$. These two quantities can be obtained by a standard method adapted to a conventional multilayer mirror (see Chapter 2).

We can illustrate the power of this approach in the case of a laminar grating, which has also been studied using the "rigorous" method (Sammar et al. 1991). The case presented in Figure 4.5 corresponds to the lamellar grating with characteristics close to those used in Sammar and André (1998): 12 Mo/C bilayers, period d = 22.5 nm, ratio $\gamma = 0.2$, grating period $D = 0.24$ mm and ratio $\Gamma = 7/8$; the radiation energy is 150 eV. The layers obtained by the "rigorous" method (Sammar et al. 1991) are almost similar to those obtained using the scattering method. For all this, the calculation time is greatly reduced with the scattering method compared to the "rigorous" method.

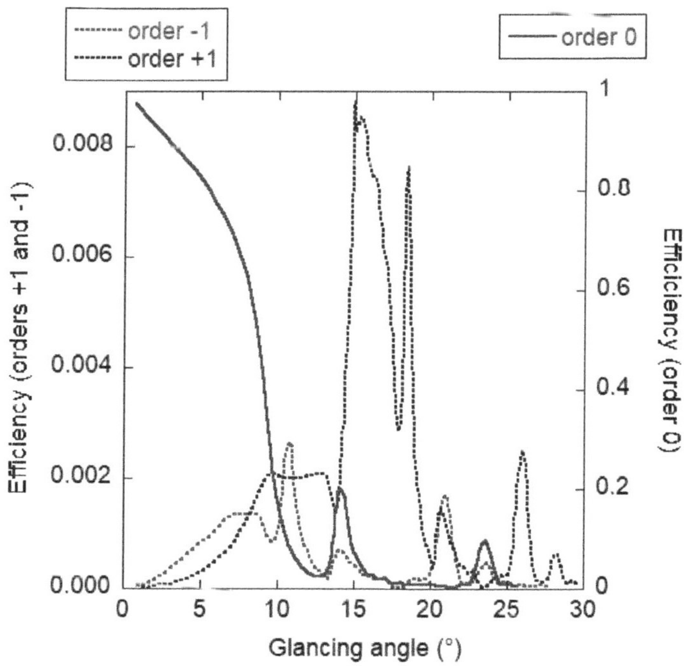

Figure 4.5. *Efficiency, for three orders of diffraction, of a lamellar grating with the characteristics given in the text. For a color version of this figure, see www.iste.co.uk/andre/xrayradiation.zip*

4.3.3. *Application to Bragg–Fresnel diffraction*

To obtain the diffraction efficiencies, according to formula [4.35] and the kinematic properties of Bragg–Fresnel lenses, the Fourier–Fresnel transform $\hat{\hat{P}}(\boldsymbol{K},r)$ of the optical potential $P_{F\pm}(x,z)$ must be calculated:

$$\widehat{\widehat{P_{F\pm}}}(\boldsymbol{K},r) = \int_{-\infty}^{+\infty} dx \int_{-\infty}^{+\infty} P_{F\pm}(x,z) \; exp\left(i\,k'\frac{(x^2+z^2)}{2\,r}\right) exp\big(i(K_x\, x + K_z\, z)\big)\, dz \qquad [4.67]$$

If the potential is given by relationship [4.9], the integral can be separated and we can write:

$$\widehat{\widehat{P_{F\pm}}}(\boldsymbol{K},r) = \widehat{\widehat{P_{F\pm x}}}(K_x)\,\widehat{\widehat{P_{F\pm z}}}(K_Z) \qquad [4.68]$$

We point out that this integral is not always separable because there are cases where the optical potential is not separable. This situation will not be considered.

In the case of P or N lenses constructed using a bilayer Bragg structure, we obtain:

$$\widehat{\widehat{P_{F\pm}}}(\boldsymbol{K},r) = \widehat{\widehat{\mathcal{F}_{\pm}}}\,(K_x)\,\hat{Z}(K_z) \qquad [4.69]$$

The transform $\widehat{\widehat{\mathcal{F}_{\pm}}}\,(K_x)$ is expressed using complex Fresnel integrals Fr[4] as follows:

$$\widehat{\widehat{\mathcal{F}_{\pm}}}\,(K_x) = \pm i\sqrt{\zeta}\; exp\left(-\,i\,\frac{\zeta}{2\pi}{K_x}^2\right)\textstyle\sum_j(-1)^{j+1}\left\{\mathrm{Fr}\left[\left(a\sqrt{j}+\frac{\zeta}{\pi}\,K_x\right)\sqrt{1/\zeta}\right]+\mathrm{Fr}\left[\left(a\sqrt{j}-\frac{\zeta}{\pi}\,K_x\right)\sqrt{1/\zeta}\right]\right\}\;;\;\zeta=\frac{\pi\,r}{k\prime} \qquad [4.70]$$

The sum on j extends from 0 to 2L+1 for P lenses and from 0 to 2L+2 for N lenses. This sum can become heavy in practice, where the value of L is greater by a few dozen. In this case, it is useful to rewrite the optical potential in the form $\widetilde{\mathcal{F}_{\pm\infty}}(x)\; rect\big[x/a\sqrt{L}\big]\; Z(z)$, where $\widetilde{\mathcal{F}_{\pm\infty}}(x)$ is relative to a lens of infinite extension $(L\to\infty)$ and where the function *rect* expresses the horizontal limit. Details of the calculations are given in Sammar and André (1993b). The Fresnel–Fourier transform

4. The Fresnel Fr integrals are defined by: $\mathrm{Fr}(w) = C(w) + i\,S(w)$ with $C(w) = \int_0^{\infty} cos\left(\frac{\pi}{2}\,t^2\right)\,dt$ and $S(w) = \int_0^{\infty} sin\left(\frac{\pi}{2}\,t^2\right)dt$.

$\widehat{\widehat{P_{F\pm z}}}$ can be confused with the Fourier transform $\widehat{P_{F\pm z}}$ that corresponds to the optical potential $Z(z)$, which is given by:

$$\widehat{P_{F\pm z}} = -\left[P_a \frac{1-exp(i\,K_z\,d_a)}{K_z} + P_b \frac{1-exp(i\,K_z\,d_a)}{K_z} exp(i\,K_z\,d_a)\right] U_{N+1}\left(K_z \frac{d}{2}\right)$$

[4.71]

Using the formalism above, the diffraction performances of a linear Bragg–Fresnel lens can be calculated, in particular the spatial distribution of the diffracted radiation. An example is given in Figure 4.6. This is presented in the plane (r,θ), where r and θ are respectively the distance from the observation point to the center of the lens, and the angle between the axis of observation and the axis of symmetry of the lens. The calculation is performed for a photon energy of 930 eV (1.3 nm) and a positive lens with the following characteristics: multilayer structure formed from 100 bilayers Mo/C, with d_{Mo} = 1.5 nm, d_C = 2.5 nm, lens a = 7.2 nm, number of lines L = 50.

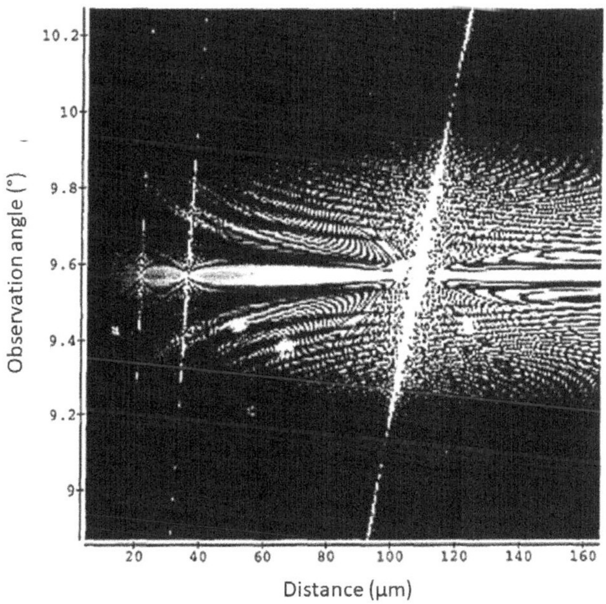

Figure 4.6. *Distribution in the (r,θ) plane of the intensity diffracted by a linear positive Bragg–Fresnel lens. The white zones indicate areas of high intensity. The radiation energy is 930 eV and the Bragg angle is 9.57°. The lens parameters are given in the text*

We note that the distribution is identical to that of an identical but negative lens (Sammar and André 1993b) in accordance with the Babinet theorem. We differentiate between the first three focal zones (1, 3, 5) formed from the centers and from a secondary distribution of complex intensity. The focal distances are given by the relationship $r_{2p+1} = \frac{a^2}{2P+1}\frac{\sin^2(\theta_B)}{\lambda}$, that is, in this case $r_1 = 108$ μm, $r_3 = 36$ μm and $r_5 = 21.6$ μm.

4.4. References

André, J.-M. and Barchewitz, R. (1994). Optique de Bragg-Fresnel multicouche pour le rayonnement X-UV. *Annales de Physique*, 19, C1-17–C1-24.

André, J.-M., Sammar, A., Kahn Malek, C., Troussel, P., Bac, S., Barchewitz, R., Pardo, B., Berrouane, H., Moreno, T., Ladan, F.R. et al. (1992). Multilayer gratings for the soft X-ray region. *Review of Scientific Instruments*, 63(1), 1399–1403.

Aristov, V.V., Erko, A.I., Martynov, V.V. (1988). Principles of Bragg–Fresnel multilayer optics. *Revue de Physique Appliquée*, 23(10), 1623–1630.

Barbee, T.W. (1989). Combined microstructure X-ray optics. *Review of Scientific Instruments*, 60(7), 1588–1595.

Barton, G. (1989). *Elements of Green's Functions and Propagation: Potentials, Diffusion, and Waves*. Oxford University Press, Oxford/New York.

Botten, L.C., Craig, M.S., McPhedran, R.C. (1981a). Highly conducting lamellar diffraction gratings. *Optica Acta: International Journal of Optics*, 28(8), 1103–1106.

Botten, L.C., Craig, M.S., McPhedran, R.C., Adams, J.L., Andrewartha, J.R. (1981b). The finitely conducting lamellar diffraction grating. *Optica Acta: International Journal of Optics*, 28(8), 1087–1102.

Bracewell, R.N. (2000). *The Fourier Transform and its Applications*. McGraw Hill, New York [Online]. Available at: http://archive.org/details/TheFourierTransformAndItsApplicationsBracewell [Accessed 30 January 2023].

Cohen-Tannoudji, C., Dupont-Roc, J., Grynberg, G. (2012). *Processus d'interaction entre photons et atomes*. EDP Sciences, Les Ullis.

Denisyuk, Y. (1963). The reflection of optical properties of an object in the wave field of its scattered radiation. *Optika I Spektroskopiya*, 15(4), 522–532.

Diveki, Z., Bourassin-Bouchet, C., de Rossi, S., English, E., Meltchakov, E., Gobert, O., Guénot, D., Carré, B., Salières, P., Ruchon, T. et al. (2014). Theoretical and experimental studies of broadband phase-controlled attosecond mirrors. *Journal of Modern Optics*, 61(2), 122–131.

Economou, E.N. (1983). *Green's Functions in Quantum Physics*. Springer Verlag, Berlin/Heidelberg [Online]. Available at: https://link.springer.com/book/10.1007/978-3-662-02369-3 [Accessed 27 January 2023].

Erko, A.I. (1990). Synthesized Bragg–Fresnel multilayer optics. *Journal of X-Ray Science and Technology*, 2(4), 297–316.

Fresnel, A.J., de Sénarmont, H.H., Verdet, E., Fresnel, L.F. (1866). *Œuvres complètes d'Augustin Fresnel*. Imprimerie impériale, Paris [Online]. Available at: http://archive.org/details/uvrescompltesda00fresgoog [Accessed 27 January 2023].

Guichet, C., Rivoira, R., Rasigni, G., Charai, A. (1994). Laterally graded division parameter layered synthetic microstructures. *Microscopy Microanalysis Microstructures*, 5(3), 189–202.

Khan Malek, C. (1991). A review of microfabrication technologies: Application to X-ray optics. *Journal of X-Ray Science and Technology*, 3(1), 45–67.

Kortright, J.B., Rice, M., Carr, R. (1995). Soft-X ray Faraday rotation at Fe L2,3 edges. *Physical Review B*, 51(15), 10240–10243.

Krastev, K., André, J.-M., Barchewitz, R. (1996). Further applications of a recursive modal method for calculating the efficiencies of X-UV multilayer gratings. *JOSA A*, 13(10), 2027–2033.

Li, L. (1993). Multilayer modal method for diffraction gratings of arbitrary profile, depth, and permittivity. *JOSA A*, 10(12), 2581–2591.

Maystre, D. (1980). Integral methods. In *Electromagnetic Theory of Gratings*, Petit, R. (ed.). Springer, Berlin/Heidelberg, 63–100. doi: 10.1007/978-3-642-81500-3_3 [Accessed 27 January 2023].

Morawe, C., Ziegler, E., Peffen, J.-C., Kozhevnikov, I.V. (2002). Design and fabrication of depth-graded X-ray multilayers. *Nuclear Instruments and Methods in Physics Research Section A: Accelerators, Spectrometers, Detectors and Associated Equipment*, 493(3), 189–198.

Morlens, A.-S., Ravet, M.-F., Laude, V., López-Martens, R., Roulliay, M., Jérome, A., Delmotte, F., Bridou, F., Kazamias, S., Balcou, P. et al. (2005). X-UV chirped mirror fabrication for phase control of attosecond pulses. *Proceedings of SPIE*, 5963, 260–269.

Nagel, D.J., Gilfrich, J.V., Barbee, T.W. (1982). Bragg diffractors with graded-thickness multilayers. *Nuclear Instruments and Methods in Physics Research*, 195(1), 63–65.

Nevière, M. (1991). Multilayer coated gratings for X-ray diffraction: Differential theory. *JOSA A*, 8(9), 1468–1473.

Petit, R. (1980). *Electromagnetic Theory of Gratings*. Springer-Verlag, Berlin/Heidelberg/New York [Online]. Available at: http://link.springer.com/book/10.1007/978-3-642-81500-3.

Puik, E.J., van der Wiel, M.J., Lambooy, P., Verhoeven, J., Christensen, F.E., Padmore, H.A. (1991). Characterization of a multilayer coated laminar reflection grating at $\lambda = 0.154$ nm. *Journal of X-Ray Science and Technology*, 3(1), 19–34.

Reuse, F.A. (2007). *Electrodynamique et optique quantiques*. EPFL Press, Lausanne [Online]. Available at: https://www.epflpress.org/produit/456/9782880747053/electrodynamique-et-optique-quantiques [Accessed 30 January 2023].

Rife, J.C., Hunter, W.R., Barbee, Jr., T.W., Cruddace, R.G. (1989). Multilayer-coated blazed grating performance in the soft X-ray region. *Applied Optics*, 28(15), 2984.

Rife, J.C., Barbee, Jr., T.W., Hunter, W.R., Cruddace, R.G. (1990). Performance of a tungsten/carbon multilayer-coated, blazed grating from 150 to 1700 eV. *Physica Scripta*, 41(4), 418.

Sammar, A. and André, J.-M. (1993a). Diffraction of multilayer gratings and zone plates in the X-ray region using the Born approximation. *JOSA A*, 10(4), 600–613.

Sammar, A. and André, J.-M. (1993b). Dynamical theory of stratified Fresnel linear zone plates. *JOSA A*, 10(11), 2324–2337.

Sammar, A. and André, J.-M. (1998). Integral calculation method of the efficiencies of laminar stratified gratings. *Optics Communications*, 149(4), 348–354.

Sammar, A., André, J.-M., Pardo, B. (1991). Diffraction and scattering by lamellar amplitude multilayer gratings in the X-UV region. *Optics Communications*, 86(2), 245–254.

Stemmler, P., Bizeuil, C., Marmoret, R., André, J.-M., Barchewitz, R., Rivoira, R. (1999). Calibration and spectroscopy with a laterally-d graded multilayer interference mirror. *Journal of Electron Spectroscopy and Related Phenomena*, 101/103, 919–925.

Troussel, P., Maury, H., Bridou, F., Delmotte, F., Rosch, R., Reverdin, C., Meltchakov, E. (2011). Miroirs multicouches apériodiques à large bande passante (2–10 keV) pour les diagnostics d'imagerie X à haute résolution spatiale. In *UVX 2010 – 10ème Colloque sur les Sources Cohérentes et Incohérentes UV, VUV et X ; Applications et Développements Récents*. EDP Sciences, Paris, 227–231 [Online]. Available at: https://uvx.edpsciences.org/articles/uvx/abs/2011/01/uvx2011_034/uvx2011_034.html [Accessed 26 January 2023].

Vincent, P. (1980). Differential methods. In *Electromagnetic Theory of Gratings*, Petit, R. (ed.). Springer, Berlin/Heidelberg, 101–121. doi: 10.1007/978-3-642-81500-3_4 [Accessed 27 January 2023].

Warburton, W.K. (1990). On the diffraction properties of multilayer coated plane gratings. *Nuclear Instruments and Methods in Physics Research Section A: Accelerators, Spectrometers, Detectors and Associated Equipment*, 291(1), 278–285.

5

Optimization and Design of Artificial Bragg Mirrors

Jean-Michel ANDRÉ and Philippe JONNARD
LCPMR, CNRS, Sorbonne Université, Paris, France

5.1. Theoretical optimization of the mirrors

5.1.1. *Introduction*

Optimization of the performance of an artificial Bragg mirror depends on the mirror's intended purpose and the spectral domain in which it is used. If the mirror is used for its selective properties, in other words as a dispersive element – for example, in spectrometric analysis to determine the composition of a material – the performance of the resolving power and the "signal-to-noise" ratio will be prioritized. If the mirror is used for its specular properties and in imaging – for example, for telescopes in astrophysics or as reflectors in lithography – we will seek to maximize the reflecting power: peak or integrated reflectance. The optimization conditions depend in particular on the spectral region concerned within the X-ray radiation domain.

The X-ray radiation domain (from 0.01 to 50 nm or from 100 keV to 25 eV) can be, more or less arbitrarily, divided into several regions:

– hard X-rays (region I) with wavelengths between 0.01 and 4 nm;

– soft X-rays between 4 and 13 nm, including the following subdivisions:

- the water window (region II) with wavelengths between 2.3 nm (oxygen K edge) and 4.4 nm (carbon K edge),

X-ray Radiation and Artificial Bragg Structures,
coordinated by Jean-Michel ANDRÉ and Philippe JONNARD.

- the carbon window (region III) with wavelengths between 4.4 nm (carbon K edge) and 5 nm,

- the extreme UV (EUV) lithography region (region IV) with wavelengths between 6 nm (boron K edge) and 13.3 nm (silicon L edge);

– the astrophysics region (region V) with wavelengths between 17 and 50 nm.

5.1.2. *Role of the number of unit cells*

Whatever the type of unit cell used (bilayer, trilayer, etc.), the number of unit cells N_{cu} has a very significant influence on the reflectance. In the absence of all attenuation (absorption, extinction, diffuse scattering), a kinematic approach shows that peak reflectance should increase according to N_{cu}^2. In practice, the attenuation effects limit the number of unit cells that are effectively involved in the diffraction process. The effective unit cells are spread over a depth known as the "extinction length". The peak reflectance tends asymptotically to a limit value less than 1 when N_{cu} tends to infinity if the Bragg resonance condition is verified, as shown in Figure 5.1.

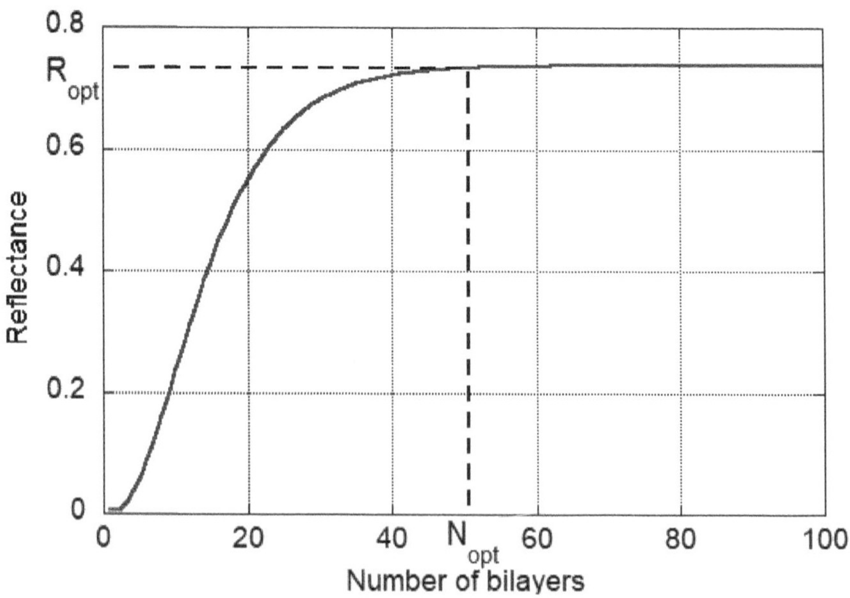

Figure 5.1. *Reflectance of an Mo/Si multilayer with a period of 6.9 nm, at normal incidence and at the 13.5 nm wavelength, as a function of the number of bilayers*

Determination of the number of unit cells required to reach a peak reflectance value close to the asymptotic value depends on the type of unit cell (bilayer or multilayer), the value of the period and the internal structure, in particular the division ratio γ[1] for bilayer unit cells.

5.1.3. *Case of a bilayer unit cell*

5.1.3.1. *Optimization in terms of reflectance*

A considerable number of theoretical (Rosenbluth 1988; Yamamoto and Namioka 1992) and experimental studies have been performed on optimization in terms of reflectance related to the choice of materials. A pair of materials that appears "theoretically" to be of interest, generally because it offers a good refractive index contrast and low radiation absorption, can turn out to be of little interest in practice because it gives rise to problems such as a high geometric roughness at the interfaces or interdiffusion phenomena that affect the reflectance. We will discuss these imperfection issues (roughness, interdiffusion, etc.) again later. V.V. Lider recently wrote a review article in which this problem of the choice of materials is discussed in detail (Lider 2019). This question was also reviewed in an older article (André and Barchewitz 2002). Among the simple rules used in making a choice of materials, we select the following:

– minimize the absorption by materials, which means that penetration of the incident radiation increases and therefore the number N_{eff} of unit cells involved in the diffraction increases; as previously seen, the peak reflectance increases with N_{eff};

– maximize the refractive index contrast, in other words the absolute value of the difference between the real parts of the optical indices;

– improve the quality of the interfaces, tending towards flat and non-diffuse interfaces, otherwise the radiation will be scattered outside the specular direction.

To satisfy the first two conditions for a given radiation of energy E, the strategy consists of choosing materials with slightly higher absorption thresholds in terms of energy at the E value. Therefore, we benefit from low absorption and from the anomaly effect on the real part of the index. The current state of the art regarding the choice of materials is given in Table 5.1.

1. The division ratio γ is the ratio of the thickness of a layer (generally the densest and thinnest) to the total thickness of the bilayer.

Region	Pair of Materials
I	W/C, W/B_4C, W/Si, Mo/C, Mo/B_4C, Mo/Si, Pt/C
II	Cr/Sc, Cr/Ti
III	Co/C, Ni/C, Fe/C, V/C
IV	Mo/Si, Mo/B, La/B, Pd/Sr, Mo/Sr, Ru/Sr
V	Al/Zr, Co/Mg, Zr/Mg

Table 5.1. *Typical pairs of materials to obtain high reflectance, according to the spectral regions in the X-ray radiation domain*

Optimization in terms of reflectance related to the geometry of the bilayer, in other words period *d* and ratio γ, is obtained from the theoretical considerations presented in Chapter 2. Concerning the value of the period *d*, the rule is briefly speaking to minimize the material thickness effectively participating in the diffraction, in order to minimize the absorption effect, therefore globally reducing the value of *d* as much as possible. The consequence of this is an increase in the Bragg angle for a given energy. However, for *TM* polarization, approaching the Brewster angle in the vicinity of 45° must be avoided because, as shown in Chapter 2, within this angular range the reflectance tends to extinguish itself with this polarization. If the geometry allows it, approaching a normal angle of incidence is a good strategy because the geometry known as *backscattering*, in other words the Bragg angle close to 90°, increases the angular acceptance of a Bragg reflector (Kushnir and Suvorov 1990).

In fact, the influence of absorption is complex and requires deeper analysis. Optimization in terms of maximum reflectance (in other words with the aim of obtaining a peak or integrated reflectance) requires indissociable involvement of the type of materials, the ratio γ and the period *d* in addition to the number of bilayers, as we will show hereafter. Once the nature of the materials is fixed, the choice of the ratio γ in order to obtain the optimal value γ_{opt} is guided by the following relationship obtained from different theoretical approaches (Kozhevnikov and Vinogradov 1987; Pardo et al. 1988):

$$tan(\pi\,\gamma_{opt}) = \pi\,\gamma_{opt} + W\,; W = \pi\,\frac{\beta_b}{\beta_a-\beta_b} \qquad [5.1]$$

A demonstration of this relationship is given in Appendix 1. The relationship is a transcendental equation that makes the imaginary parts β_a and β_b of the optical index of materials **a** and **b** of the bilayer apparent. Rosenbluth (1988) proposed an iterative method for resolving equation [5.1]:

$$\text{If} < 0{,}66 : \gamma'_0 = \sqrt[3]{3\,W - 2.5\left(\frac{W}{1+W}\right)^{5/3}}\ ; \gamma'_{I+1} = \gamma'_I - \frac{tan(\gamma'_I)-\gamma_I-W}{sec^2(\gamma'_I)-1}$$

$$\text{If} > 0{,}66 : \gamma'_0 = \frac{1}{2}\left[\pi - \left(W+\frac{\pi}{2}\right) + \sqrt{\left(W+\frac{\pi}{2}\right)^2 - 4}\right];$$

$$\gamma'_{I+1} = Arctan\,(\gamma'_I + W) \qquad [5.2]$$

where $\gamma' = \pi\,\gamma$, and the index indicates the level of iteration. In practice, it appears that to optimize the first order of diffraction with a small number of bilayers and a low absorption of the materials, the optimal value of γ_{opt} is close to 1/2.

The optimal value d_{opt} of the period and the number N_{opt} of bilayers required to reach a value close to the asymptotic reflectance can be obtained using the coupled-wave theory. The strategy consists of considering the limit case corresponding to $N \to \infty$ of the limit reflection coefficient written as r_∞. The limit reflectance denoted as R_∞ is obtained using relationship [3.21] and is given by:

$$R_\infty(\tilde{\eta}) = |r_\infty|^2 \equiv \lim_{N\to\infty}\left|\frac{\tilde{K}}{\tilde{\alpha}+\sqrt{\tilde{\alpha}^2-\tilde{K}^2}}\right|^2 = \left|\frac{\tilde{\eta}}{1+\sqrt{1-\tilde{\eta}^2}}\right|^2 \qquad [5.3]$$

where we have stated:

$$\tilde{\eta}(\mathrm{d}) \equiv \frac{\tilde{K}}{\tilde{\alpha}} = \frac{1-i\,t(n^*_a,n^*_b)}{1-i\,\rho(d,\gamma)}\,A(\gamma,n^*_a,n^*_b) \qquad [5.4]$$

$\tilde{\eta}$ is a quantity that depends on the period d via $\rho(d,\gamma)$. This latter quantity $\rho(d,\gamma)$, not given explicitly, generalizes the case used by Vinogradov and Zeldovich (1977) for normal incidence to the case of an oblique angle of incidence. The quantities $t(n^*_a,n^*_b)$ and $A(\gamma,n^*_a,n^*_b)$ are given by:

$$t(n^*_a,n^*_b) = \frac{Re[(n^*_a)^2-(n^*_b)^2]}{Im[(n^*_a)^2-(n^*_b)^2]},\ A(\gamma,n^*_a,n^*_b) = \frac{sinc(\gamma\,\pi)}{\left(1+\frac{\gamma^{-1}\,Im[(n^*_b)^2]}{Im[(n^*_a)^2-(n^*_b)^2]}\right)} \qquad [5.5]$$

The optimal period d_{opt} is obtained by differentiating R_∞ with respect to d via $\tilde{\eta}(d)$ and by seeking the condition to be fulfilled to verify $Re\left(\frac{\partial R_\infty(\tilde{\eta})}{\partial\tilde{\eta}}\right) = 0$.

Following a relatively onerous calculation inspired by the works of Vinogradov and Zeldovich (1977), this condition results in:

$$R.Im\left[(1-i\,\rho)\left(1-\left(\tilde{\eta}(d_{opt})\right)^2\right)\right]=0 \quad [5.6]$$

that is, expressing $\tilde{\eta}(d_{opt})$ using equation [5.5]:

$$\rho(d_{opt}) = A^2\,t \quad [5.7]$$

from which the value of d_{opt} is deduced. The values of γ_{opt} and d_{opt} can be used to obtain the optimal value of limit reflectance $R_{opt\infty}$:

$$R_{opt\infty} = \frac{a^2(1+t^2)}{a^4t^2\left(1+\sqrt{(1-a^2)(1+a^2t^2)}\right)^2}\;;a=cos(\pi\,\gamma_{opt}) \quad [5.8]$$

The number of bilayers that can also be qualified as optimal and denoted as N_{opt} is given by:

$$N_{opt}\sim\frac{\sin(\theta)}{d_{opt}\,\bar{\mu}} \quad [5.9]$$

where $\bar{\mu}$ is the lineic attenuation coefficient (of dimension $[L^{-1}]$) of a bilayer and θ is the glancing angle. The quantity N_{opt} in fact gives the required number of deposited bilayers in order to achieve a value "close" to $R_{opt\infty}$. The notion of "close" is of course subjective, and an increase in the number of bilayers with respect to N_{opt} will lead at least theoretically to a slight increase in the reflectance. However, it has been observed that in many cases, an increase in the number of layers deposited does not in practice improve the reflectance or lead to deterioration of it following the appearance of defects in the structure (Guichet et al. 1997; Ghafoor et al. 2006).

5.1.3.2. *Optimization in terms of bandwidth*

The strategy for reducing the bandwidth is based on several approaches. If we consider the *ideal case of a non-absorbing medium*, the relative bandwidth $\mathcal{R}=\frac{\Delta E}{E}=\frac{\Delta\lambda}{\lambda}$ is given with the usual notations (Lider 2019) by:

$$\mathcal{R}=\frac{\sqrt{\left(1+C^2\,y^2\left(\frac{Re(\Delta\chi)}{Im(\Delta\chi)}\right)\right)}}{(1-C^2\,y^2)}\frac{Im(\bar{\chi})}{sin^2(\theta)}\;;y=\gamma\,\frac{sinc(p\,\gamma\,\pi)}{(\gamma+\frac{Re(\chi_b)}{Im(\Delta\chi)}} \quad [5.10]$$

where C is a polarization factor. Relationship [5.10] is reduced in the *ideal case of a non-absorbing medium* to:

$$\mathcal{R} = |Re(\Delta\chi)|\, \gamma\, \frac{sinc(p\, \gamma\, \pi)}{sin^2(\theta)} \qquad [5.11]$$

Using this formula, it follows that to reduce the value of $\mathcal{R}$, $|Re(\Delta\chi)|$ must be reduced, which is problematic because in this case, this operation causes a drop in reflectance (which is proportional to the square of this quantity, again for the case of non-absorbing media). However, in the field of hard X-ray radiation where absorption remains low, this approach, which consists of combining materials with low optical index contrast and a high number of bilayers, is useful in reducing the value of $\mathcal{R}$: thus, Martynov et al. (2004) used the pair alumina (Al_2O_3)/boron carbide (B_4C) to obtain a value of $\mathcal{R}$ equal to 2.7 10^{-4} and a reflectance of 40% with a multilayer structure comprising 800 bilayers with thickness 2.6 nm.

Another route consists of reducing the value of γ, which means reducing the relative thickness of the type **a** layer at the risk of moving away from the optimal value for the reflectance γ_{opt} given by equation [5.1] for a real, therefore absorbing, medium; for a value of γ below 0.2, the reflectance drops considerably. Working at a higher order, so with $p > 1$, the value of $\mathcal{R}$ can still be reduced using the function $sinc(p\, \gamma\, \pi)$ but again to the detriment of the reflectance. Therefore, Yulin et al. (2005) obtained a value of $\mathcal{R} \approx 6\; 10^{-3}$ at the wavelength 13.5 nm. Generally, for these last approaches, a compromise must be found between the narrow bandwidth and sufficient reflectance. Operating at a high Bragg angle is also important in order to reduce the value of $\mathcal{R}$ which varies according to $sin^{-2}(\theta)$. Using the so-called *backscattering* geometry, the effects of angular broadening $\Delta\theta$ associated with the width of the Darwin diffraction pattern can also be canceled. Indeed, by differentiating the Bragg law, we obtain:

$$\frac{\Delta\lambda}{\lambda} = \frac{\Delta d}{d} + cotan\theta\, \Delta\theta \qquad [5.12]$$

If we assume that there is no error on the period ($\Delta d = 0$), then $\frac{\Delta\lambda}{\lambda} \rightarrow 0$ when $\theta \rightarrow \frac{\pi}{2}$.

In fact, in the soft X-ray radiation domain, the principal strategy for a real diffracting (therefore absorbing) medium is based on a reduction of attenuation of the radiation penetrating into the diffracting medium and therefore an increase in the number of unit cells, in this case of bilayers, participating in diffraction. Selection of materials with low absorbing levels will therefore be prioritized and, more generally, selection of a lighter medium. To do so, a method that has given promising results is the use of lamellar diffraction gratings (see Figure 3.1a), with optimized parameters. The principle of this method was given by Sammar et al. (1993) and was confirmed theoretically and experimentally in a series of works (André et al. 2002; Benbalagh et al. 2005; Jonnard et al. 2009; Kozhevnikov et al. 2010). The performances of the narrow bandwidth monochromators will be described in more detail in Chapter 7. At this stage, we can summarize the performances of these devices as follows: theoretically, in the specular reflection conditions with an angular divergence $\Delta\psi$ and in "mono-mode" regime, which is satisfied if $\Gamma D \Delta\psi \ll d$ (with the usual notations), the spectral resolution $\mathcal{R}$ can be reduced by a factor Γ while maintaining a reflectance close to the reflectance of a conventional mirror (maintained to the order of 80%). In practice, in the soft X-ray radiation domain, the various defects introduced by the fabrication techniques have not enabled the ultimate theoretical performances in terms of reflectance to be achieved. Another reason why these gratings are useful for the spectroscopic analysis is to reduce the "specular component" of the reflection (Bragg diffraction at order 0) and thus to improve the signal/noise ratio (Fialin et al. 1996).

If, on the contrary, we wish to increase the bandwidth to obtain so-called "super-mirrors" which work in the hard X-ray radiation domain under a grazing angle of incidence, and which were developed in the 1990s (Joensen et al. 1994, 1995; Yamashita et al. 1998), the density of the materials must be increased and the periodicity set aside in order to create so-called aperiodic structures. These super-mirrors, which generally have a period gradient, were then significantly developed in astrophysics for telescopes or synchrotron lines operating under a grazing incidence for hard X-ray radiation. They have also been developed in the soft X-ray and EUV domain to operate generally under quasi-normal incidence (Pirozhkov and Ragozin 2015; Barysheva et al. 2019).

A comparable situation occurs for "chirped" mirrors used to reflect very short pulses of soft X-ray radiation (lasting for the order of an attosecond) which have a wide spectrum due to the shortness of the pulse (Heisenberg principle) (Bourassin-Bouchet et al. 2011). This type of mirror is aperiodic with a layer distribution profile which also depends on the constraints on the phase, as we will discuss in section 5.3 regarding design. In both cases (super-mirrors and chirped mirrors), determination of the optimal profile is a difficult problem which is generally based on optimization algorithms.

5.1.3.3. *Case of a multilayer unit cell*

Mirrors constructed with a trilayer or quadri-layer unit cell were first created in 1990 (Boher et al. 1990). In this work, it was demonstrated that by using a *multilayer* unit cell, the reflectance can be improved with respect to a bilayer system and material selection criteria are given. The usefulness of this approach has been confirmed in the spectral region of wavelengths between 11 and 14 nm, covering the spectral region of EUV lithography (Singh and Braat 2000).

In fact, it is possible to demonstrate that under certain conditions, the reflectance of a *trilayer (or multilayer)* unit cell obtained from a *bilayer unit cell* by adjoining one (or several) layer(s) can be increased so it is greater than the reflectance of the generating bilayer unit cell.

Generally, the foundation for this approach is based on optimizing the distribution of the optical index within the unit cell. In 1980, Cariglia and Apfel (1980) took this route, recommending that the thickness of the densest (absorbing) material within the unit cell should be modified: the thickness of the unit cell d is kept constant along the stack, but the thickness of the dense layer d_{abs} increases nearer the substrate, or in other words the ratio $\gamma = \frac{d_{abs}}{d}$ increases with depth.

Optimization in terms of reflectance related to the choice of materials for a multilayer unit cell is even more complicated than for bilayer unit cells. The problem was studied by Larruquert (2001, 2002a, 2002b) in a series of articles. This author focused primarily on the spectral domain corresponding to wavelengths longer than 30 nm and gave a certain number of criteria that can be represented on a graph, forming a polygon in the $\{Re(n^*), \beta = Im(n^*)\}$ plane such as in Figure 5.2.

An important rule when selecting materials is written using the optical index increment, in other words the difference in index between two successive layers $\Delta n^*_i = n^*_{i+1} - n^*_i$. The criterion is written as:

$$Im\left[\frac{\Delta n^*_i}{\Delta n^*_{i-1}}\right] < 0 \qquad [5.13]$$

which means that we are turning in the complex plane of indices, in the clockwise direction and by an angle less than 180°, as represented by the arrow in Figure 5.2. The materials and the order of their deposition are selected by moving along the minimal polygon in the clockwise direction following the arrows.

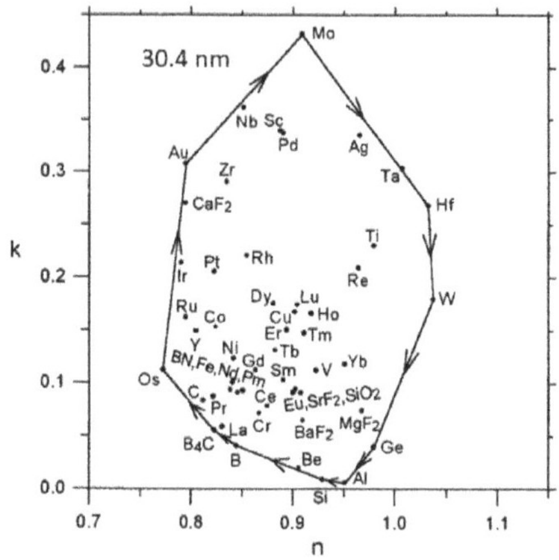

Figure 5.2. *Optical constants of various materials at the wavelength of 30.4 nm. The contour with arrows is the smallest polygon including all the materials. Reproduced with permission from Larruquert (2002a)*[2]*. © The Optical Society*

5.2. Practical optimization of mirrors

The presence of a series of defects is inherent to the fabrication of these artificial structures (deposition technique, etching technique, etc.) and is also the result of natural phenomena (crystallization, interdiffusion, etc.). As a result, the performances, especially in terms of reflectance, are lower than predicted by the theory for ideal structures, considered exempt from defects. We will consider some of these defects and briefly describe some techniques implemented to reduce their effects.

5.2.1. *Accidental defects*

Accidental defects that we will consider are geometrical. They may be numerous and varied. Among the most serious for a multilayer structure, there is the absence or appearance of one or several layers due to technical problems during deposition. There are also errors in the thickness of the layers that are related to the continuous drift on certain parameters (temperature, pressure, etc.) during fabrication, or modifications to the thickness of the layers associated with random variations in the deposition conditions. The list is not exhaustive. As examples, we can describe two modelings of this type of defect. The first assumes a linear drift, whereas the second

2. See doi: 10.1364/JOSAA.19.000391.

assumes a random but limited variation of the position of the layers associated with a modification of the scattering power.

Let us consider a linear drift (Karpov and Stolyarov 1993). The dielectric constant, assumed to be real, is written in the form:

$$\varepsilon(z) = \bar{\varepsilon} + \sum_m^{m\neq 0} \varepsilon_m \, e^{i\,g\,m\,z} + \xi\, z \,, m \in \mathbb{Z}^* \qquad [5.14]$$

where ξ is the linear drift coefficient with a dimension of inverse length verifying $\xi\, L \ll \bar{\varepsilon}$, where L is the total thickness of the structure. Using coupled-wave theory (see Chapter 3), the amplitudes of the direct $\mathcal{F}(z)$ and retrograde $\mathcal{R}(z)$ waves are written as:

$$\frac{d}{dz}\begin{pmatrix}\mathcal{F}(z)\\ \mathcal{R}(z)\end{pmatrix} = i\begin{pmatrix} 0 & \varkappa_n\, e^{(\alpha-2i\delta)\,z}\, e^{-i\frac{\sigma}{2}z^2} \\ -\varkappa_{-n}\, e^{-(\alpha-2i\delta)\,z}\, e^{i\frac{\sigma}{2}z^2} & 0 \end{pmatrix}\begin{pmatrix}\mathcal{F}(z)\\ \mathcal{R}(z)\end{pmatrix} \qquad [5.15]$$

with $\sigma = \frac{k_0}{\sqrt{\bar{\varepsilon}}}\,\xi$ with a dimension inverse to the square root length. The coupling coefficients $\varkappa_{\pm n}$ are from Kogelnik's coupled-wave theory, with phase shift of $\alpha - 2i\delta$.

The amplitudes can be written using the Weber–Hermite parabolic cylindrical functions $V^{(\pm,\pm,\pm)} = V\left(\pm i\,\rho\, \pm \frac{1}{2}, z\,\sqrt{\sigma}\, e^{\pm i\frac{\pi}{4}}\right)$ with $\rho = \frac{|\varkappa_{\pm n}|}{\sqrt{\sigma}}$, as follows:

$$\mathcal{F}(z) = \left(C_1\, V^{(+,-,-)} + C_2\, V^{(-,+,+)}\right) e^{-i\frac{\sigma}{2}z^2}$$

$$\mathcal{R}(z) = \frac{\varkappa_{-n}}{\sqrt{\sigma}}\left(C_1\, V^{(+,+,-)} + C_2\, V^{(-,-,+)}\right) e^{i\frac{\sigma}{2}z^2}\, e^{i\,3\frac{\pi}{4}} \qquad [5.16]$$

where the constants C_1 and C_2 are given by the limit conditions. The Bragg peak reflectivity R_∞ of a semi-infinite structure of this kind is then given by:

$$R_\infty \equiv \left|\frac{\mathcal{R}(0)}{\mathcal{F}(0)}\right|^2 = tanh\left(\frac{\pi}{2}\,\rho^2\right) \qquad [5.17]$$

This equation shows that when $\sigma \gg \frac{\pi}{2}\,\left|\varkappa_{\pm n}\right|^2$, the reflectivity tends to cancel out but remains high ($0.76 \approx th(1)$), if $\sigma \approx \frac{\pi}{2}\,\left|\varkappa_{\pm n}\right|$.

More generally, the influence of the variation of the thickness or index of a layer referenced by the index j with respect to a nominal value can be estimated from a differentiation calculation. We can start with the method of characteristic matrices (see Chapter 2) and follow the approach of Furman and Tikhonravov (1992). We recall that the transmission and reflection coefficients of the global characteristic matrix $\widehat{C}_g$ are given respectively by:

$$t_N = \frac{2\,\xi_{inc}}{\xi_{inc}\,c_{g11}+\xi_{subs}\,c_{g22}+\xi_{inc}\,\xi_{subs}\,c_{g12}+c_{g21}}$$

$$r_N = \frac{\xi_{inc}\,c_{g11}-\xi_{subs}\,c_{g22}+\xi_{inc}\,\xi_{subs}\,c_{g12}-c_{g21}}{\xi_{inc}\,c_{g11}+\xi_{subs}\,c_{g22}+\xi_{inc}\,\xi_{subs}\,c_{g12}+c_{g21}} \qquad [5.18]$$

where c_{gij} are the elements of $\hat{C}_g$.

Let us consider a variation in the thickness of layer j of thickness d_j. Differentiation of the transmission coefficient t_N with respect to d_j is given by:

$$\frac{\partial t_N}{\partial d_j} = -\frac{{t_N}^2}{2\,\xi_{inc}}\left[\xi_{inc}\frac{\partial c_{g11}}{\partial d_j} + \xi_{subs}\frac{\partial c_{g22}}{\partial d_j} + \xi_{inc}\,\xi_{subs}\frac{\partial c_{g12}}{\partial d_j} + \frac{\partial c_{g21}}{\partial d_j}\right] \qquad [5.19]$$

and of the reflection coefficient r_N with respect to d_j is given by:

$$\frac{\partial r_N}{\partial d_j} = \frac{t_N}{2\,\xi_{inc}}\left[\xi_{inc}\frac{\partial c_{g11}}{\partial d_j}(1-r_N) - \xi_{subs}\frac{\partial c_{g22}}{\partial d_j}(1+r_N) + \right.$$
$$\left. \xi_{inc}\,\xi_{subs}\frac{\partial c_{g12}}{\partial d_j}(1-r_N) - \frac{\partial c_{g21}}{\partial d_j}(1+r_N)\right] \qquad [5.20]$$

The global characteristic matrix $\widehat{C}_g$ of a one-dimensional artificial Bragg structure (1D-PC), composed of N unit cells, is given by the product of the N characteristic matrices $\widehat{M_j}$ that correspond to each unit cell j, so:

$$\widehat{C}_g = \prod_{j=1}^{N} \widehat{M_j} \qquad [5.21]$$

The derivative of the elements $c_{g11}, c_{g12}, c_{g21}$ and c_{g22} is obtained from the derivative $\frac{\partial \widehat{C}_g}{\partial d_j}$:

$$\frac{\partial \widehat{C}_g}{\partial d_j} = \widehat{M_1} \ldots \frac{\partial \widehat{M_j}}{\partial d_j} \ldots \widehat{M_N} \qquad [5.22]$$

Introducing the following two matrices:

$$\hat{\Gamma} \equiv -\frac{{t_N}^2}{2\,\xi_{inc}}\begin{pmatrix} \xi_{inc} & 1 \\ \xi_{inc}\,\xi_{subs} & \xi_{subs} \end{pmatrix}$$

$$\hat{\Xi} \equiv \frac{t_N}{2\,\xi_{inc}}\begin{pmatrix} \xi_{inc}(1-r_N) & -(1+r_N) \\ \xi_{inc}\,\xi_{subs}\,(1-r_N) & -\,\xi_{subs}\,(1+r_N) \end{pmatrix} \quad [5.23]$$

we can rewrite $\frac{\partial t_N}{\partial d_j}$ and $\frac{\partial r_N}{\partial d_j}$ as follows:

$$\frac{\partial t_N}{\partial d_j} = Trace\left[\frac{\partial \widehat{C}_g}{\partial d_j}\,\hat{\Gamma}\right]$$

$$\frac{\partial r_N}{\partial d_j} = Trace\left[\frac{\partial \widehat{C}_g}{\partial d_j}\,\hat{\Xi}\right] \quad [5.24]$$

Using a similar calculation, the derivative with respect to the refraction index can be obtained. If we write δp_j the variation of the parameter p_j (thickness or real part of the index) corresponding to the layer *j*, then the corresponding variations of transmittance $\delta|t_N|^2$ and reflectance $\delta|r_N|^2$ are given respectively by:

$$\delta|t_N|^2 = 2\,\frac{\xi_{subs}}{\xi_{inc}}\,\delta p_j Re\left[t_N{}^*\,\frac{\partial t_N}{\partial p_j}\right]$$

$$\delta|r_N|^2 = 2\,\delta p_j Re\left[r_N{}^*\,\frac{\partial r_N}{\partial p_j}\right] \quad [5.25]$$

For the imaginary part of the index, we replace the real part *Re* by the imaginary part *Im* in the above relationships. Calculation of higher-order derivatives can be performed by generalizing the above calculation. In fact, the errors are generally correlated, and their combined influences can be expressed in a Taylor series expansion. Therefore, for the transmittance *T*, the difference ΔT with respect to the nominal value $T_0 = |t_N|^2{}_0$, is given for non-absorbing media:

$$\Delta T = T - T_0 \approx \sum_{j=1}^{m}\left(\xi_j\,\frac{\partial T}{\partial d_j} + \varsigma_j\,\frac{\partial T}{\partial n_j}\right) + \frac{1}{2}\sum_{l,j-1}^{m}\left(\xi_i\,\xi_j\,\frac{\partial^2 T}{\partial d_\iota \partial d_j} + \varsigma_i\,\varsigma_j\,\frac{\partial^2 T}{\partial n_\iota \partial n} + 2\,\xi_i\,\varsigma_j\,\frac{\partial^2 T}{\partial d_\iota \partial n_j}\right) \quad [5.26]$$

The quantity *m* is the space dimension of the parameters (thickness and refraction index). This type of relationship is directly usable when the values of the respective differences on the thickness and indices ξ_j, ς_j are known, which is rarely the case. In practice, we most often have access to a statistical distribution of these values. Generally, the distribution law is normal (Gaussian distribution function) with a known value of the standard deviations $\sigma\xi$ and $\sigma\varsigma$. It is then possible to determine a statistical average value of the difference $\overline{\Delta T}$:

$$\overline{\Delta T} = \sum_{i=1}^{m} \left(\sigma\xi_i{}^2 \frac{\partial^2 T}{\partial d_i{}^2} + \sigma\varsigma_i{}^2 \frac{\partial^2 T}{\partial n_i{}^2} \right) \quad [5.27]$$

and the standard deviation S:

$$S = \sum_{i=1}^{m} \left(\sigma\xi_i{}^2 \frac{\partial T}{\partial d_i} + \sigma\varsigma_i{}^2 \frac{\partial T}{\partial n_i} \right)^2 + 2\sum_{i,j=1}^{m} \left(\sigma\xi_i{}^2 \sigma\xi_j{}^2 \frac{\partial^2 T}{\partial d_i \partial d_j} + \sigma\varsigma_i{}^2 \sigma\varsigma_j{}^2 \frac{\partial^2 T}{\partial n_i \partial n_j} + 2\,\sigma\xi_i{}^2 \sigma\varsigma_j{}^2 \frac{\partial^2 T}{\partial d_i \partial n_j} \right) \quad [5.28]$$

5.2.2. "Natural" defects – Rough or diffuse interfaces

Artificial stratified structures generally present imperfect interfaces as a result of geometric roughness and/or interdiffusion, with a sometimes subtle distinction between these two types of imperfection. The geometric roughness associated with an interdiffusion is represented in Figure 5.3. It is primarily related to the method of fabrication of the structure, to the chemical nature of the materials and to the quality of any substrate present.

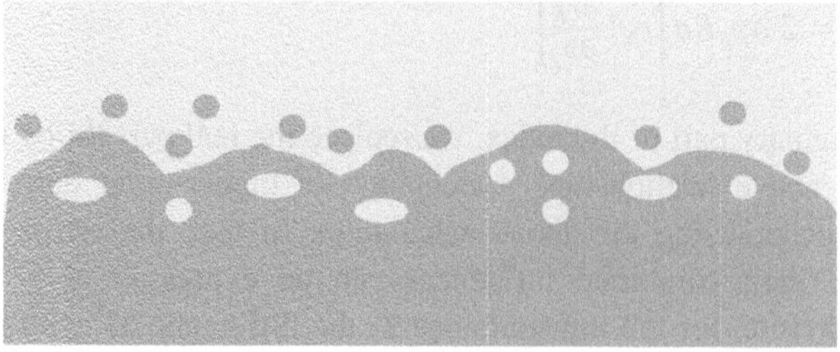

Figure 5.3. *Sketch of a rough and diffuse interface (André 2023). For a color version of this figure, see www.iste.co.uk/andre/xrayradiation.zip*

At each interface, the roughness is characterized statistically by a distribution of asperity heights. The primary parameter of this distribution, generally considered to be normal (Gaussian), is the standard deviation denoted as σ. In addition to the statistic, a horizontal correlation function (in the stratification plane) $C_{\parallel}(\xi_{\parallel})$ is often considered, with the parameter of a horizontal correlation length $\xi_{\parallel}$. An important statistical quantity is the power spectral density $W(q_{\parallel})$ which is the two-dimensional Fourier transform on the stratification plane of the function $C_{\parallel}(\xi_{\parallel})$. This quantity has the advantage of being accessible via a variety of techniques (topographical, profilometric, etc.) (see Bennett (1976) and Rasigni et al. (1981)) with results that can be compared to optical scattering measurements (Alehyane et al. 1989).

A possible correlation between the roughnesses of the different interfaces can also be modeled by a longitudinal correlation function (perpendicular to the plane of stratification) $C_\perp(\xi_\perp)$ using the parameter of a vertical correlation $\xi_\perp$ length. The presence of roughness breaks the translation symmetry in the stratification plane, which gives rise to diffuse scattering outside the specular direction.

The interdiffusion primarily associated with the physics and chemistry of the components of the layers is modeled from an optical point of view by an index profile which reflects a chemical composition profile. These imperfections cause a reduction in the reflecting power of interferential mirrors based on these stratified structures, mainly in the X-ray radiation domain where they have been widely developed. To report on this phenomenon and model the diffuse scattering, theories based on statistical models of these imperfections previously mentioned have been developed. They are generally developed in the context of Born approximation where the multiple diffusion effects are neglected or of the more elaborate *distorted-wave Born approximation* (DWBA) theory (Sinha et al. 1988; Daillant and Bélorgey 1992; Holý and Baumbach 1994; de Boer 1995). Although they remain approximate, these theories provide interesting results, at least from a qualitative point of view. Theories of this type demonstrate that the incident radiation on a stratified structure with rough and diffuse interfaces is:

– reflected specularly, in other words according to the Snell–Descartes law. This phenomenon corresponds to what is sometimes known as coherent scattering. The reflection undergoes an attenuation with respect to the "ideal" reflectance described by a multiplication coefficient *S* which depends on the statistical parameters input into the model. This coefficient is generally close to the so-called Debye–Waller factor $exp\left(-4\,{k_\perp}^2\,\sigma^2\right)$ from an eponymous and extensively used model (see hereafter) but can deviate from this depending on the type of horizontal correlation (André 1984);

– scattered outside the specular direction (incoherent scattering) with a differential effective cross-section $\frac{d\sigma_e}{d\Omega}$ $(d\Omega)$ which is the solid angle differential element with an onerous mathematical expression summarized by the form:

$$\frac{d\sigma_e}{d\Omega} \sim k^4 \sum_{i,j} (\Delta\varepsilon)_{j,j+1}\,(\Delta\varepsilon)_{i,i+1}\left(t_j + r_j\right)\left(\tilde{t_j} + \tilde{r_j}\right)\left(t_i + r_i\right)\left(\tilde{t_i} + \tilde{r_i}\right) W(Q_\parallel)\, exp\left(-i\, Q_\parallel\, (\Delta z)_{ij}\, (C_\perp)_{ij}\right) \quad [5.29]$$

Sums over the indices *I* and *j* extend through all layers of the stratified structure; $(\Delta z)_{ij}$ is the distance between the layer with index *j* and the layer with index *i*; $\Delta\varepsilon$ denotes the index contrast between the layers *j*, *j+1* or *i*, *i+1*; *t* and *r* denote the transmission and reflection coefficients at the angle of incidence respectively, while $\tilde{t}$ and $\tilde{r}$ correspond to the transmission and reflection coefficients at the angle of diffusion respectively; and $(C_{\perp})_{ij}$ is the longitudinal correlation factor between the interfaces *i* and *j*.

In the short-wavelength domain (X-ray and X-UV radiation) beyond the critical angle, the reflection terms are negligible, such that the terms $(t_j + r_j)(\tilde{t_j} + \tilde{r_j})(t_i + r_i)(\tilde{t_i} + \tilde{r_i})$ can be replaced by the product of the transmittance terms T and $\tilde{T}$. Therefore, it is possible to obtain for the effective cross-section, in the spectral domain, an expression of the type given in various works (Sinha 1996; Gibaud and Hazra 2000). Thanks to the terms T and $\tilde{T}$, specific phenomena observable with hard X-ray radiation can be reported, such as the Yoneda or "angel wings" effect (Yoneda 1963), while the term $W(q_{\parallel})$ conditions the spatial distribution pattern of the scattered radiation: overall, long horizontal correlation lengths $\xi_{\parallel}$ lead to a concentration of the scattered radiation in the specular direction while short correlation lengths result in spatial spread of the scattered radiation. A typical spatial distribution of the scattered X radiation is represented using a map in the $(Q_{\parallel}, Q_{\perp})$ plane (*Q-mapping*) in Figure 5.4. Measurement of this type of spatial distribution constitutes the basis for investigation methodologies of stratified structure defects (Salditt et al. 1996; Freitag and Clemens 2000; Störmer et al. 2007; Haase et al. 2017).

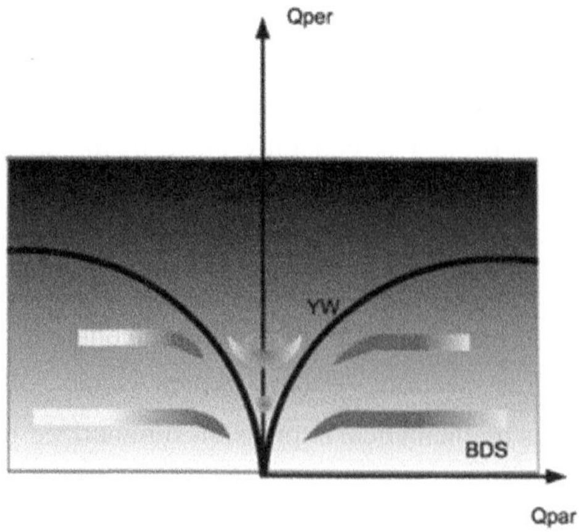

Figure 5.4. *Distribution pattern of X-ray radiation scattered by a stratified structure with rough and diffuse interfaces in the* **Q** *plane* $(Q_{\parallel}, Q_{\perp})$*; BDS: first-order and second-order Bragg scattering sheets; YW: Yoneda wings (André 2023). For a color version of this figure, see www.iste.co.uk/andre/xrayradiation.zip*

This type of modeling is still sometimes too sophisticated for engineers who design interferential mirrors and a simpler model is often used: horizontal and longitudinal correlations are neglected and both types of imperfection (geometric roughness and interdiffusion) are modeled overall at each interface by a transition layer with thickness *2e* shown in Figure 5.5. We introduce this layer in the formalism of transfer matrices as follows: at the interface, we remove a layer of thickness e from medium 2 and a layer of thickness e from medium 1, and we introduce a so-called transition layer, which is homogeneous with refraction index N.

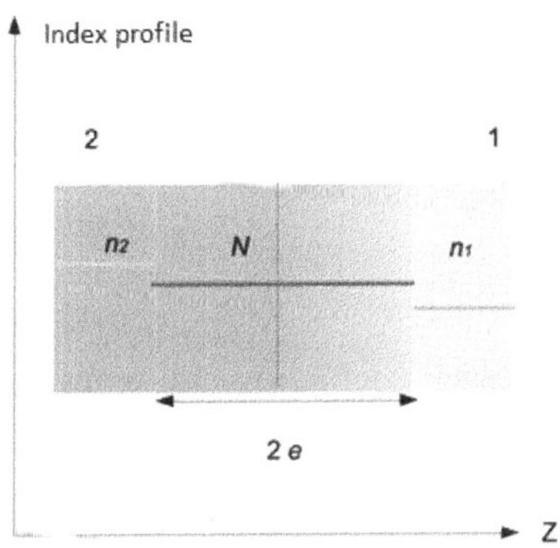

Figure 5.5. *Model of the transition layer for a rough, diffuse interface between a homogeneous medium 2 and a homogeneous medium 1; the interface with an index gradient is modeled by a transition layer of thickness 2e and index constant N, the average of the indices n_1 and n_2 (André 2023). For a color version of this figure, see www.iste.co.uk/andre/xrayradiation.zip*

The propagation matrix for the ideal interface is therefore replaced by the product of matrices:

$$\begin{pmatrix} cos(k_{\perp 2}\,2e) & \frac{-i\,k}{k_{\perp 2}} sin(k_{\perp 2}\,2e) \\ \frac{-i\,k_{\perp 2}}{k} sin(k_{\perp 2}\,2e) & cos(k_{\perp 2}\,2e) \end{pmatrix} \begin{pmatrix} cos(k_{\perp}\,2e) & \frac{i\,k}{k_{\perp}} sin(k_{\perp}\,2e) \\ \frac{i\,k_{\perp}}{k} sin(k_{\perp}\,2e) & cos(k_{\perp}\,2e) \end{pmatrix}$$

$$\begin{pmatrix} cos(k_{\perp 1}\,2e) & \frac{-i\,k}{k_{\perp 1}} sin(k_{\perp 1}\,2e) \\ \frac{-i\,k_{\perp 1}}{k} sin(k_{\perp 1}\,2e) & cos(k_{\perp 1}\,2e) \end{pmatrix} \qquad [5.30]$$

If we consider that the values of $k_{\perp 2}\,2e, k_{\perp}$ and $k_{\perp 1}\,2e$ are sufficiently low, we can calculate the matrix product with an expansion limited to the third order and, considering the dielectric constant N_t of the transition layer as the average of the dielectric constants of media 1 and 2, respectively, that is, $N = \frac{1}{4}\sqrt{n_1{}^2 + n_2{}^2}$, then the propagation matrix in the transition layer is written as:

$$\begin{pmatrix} 1 - \frac{e2}{2}\left(k_{\perp 2}{}^2 - k_{\perp 1}{}^2\right) & 0 \\ 0 & 1 + \frac{e2}{2}\left(k_{\perp 2}{}^2 - k_{\perp 1}{}^2\right) \end{pmatrix} \quad [5.31]$$

The matrix corresponding to the passages at the interfaces of the transition layer and to propagation inside it is given by:

$$\frac{1}{2}\begin{pmatrix} 1 & \frac{k}{k_{\perp 2}} \\ 1 & \frac{-k}{k_{\perp 2}} \end{pmatrix}\begin{pmatrix} 1 - \frac{e2}{2}\left(k_{\perp 2}{}^2 - k_{\perp 1}{}^2\right) & 0 \\ 0 & 1 + \frac{e2}{2}\left(k_{\perp 2}{}^2 - k_{\perp 1}{}^2\right) \end{pmatrix}\begin{pmatrix} 1 & 1 \\ \frac{k}{k_{\perp 1}} & \frac{-k}{k_{\perp 1}} \end{pmatrix} \quad [5.32]$$

If we introduce the Fresnel coefficients in terms of reflection $r_{2,1}$ and transmission $t_{2,1}$, the matrix product is given by the equation:

$$\frac{1}{t_{2,1}\, exp\left(\frac{e^2}{2}(k_{\perp 2} - k_{\perp 1})^2\right)}\begin{pmatrix} 1 & r_{2,1} exp(-\,2\, e^2\, k_{\perp 1} k_{\perp 2}) \\ r_{2,1}\, exp(-\,2\, e^2\, k_{\perp 1} k_{\perp 2}) & 1 \end{pmatrix} \quad [5.33]$$

Using this model, often qualified as Debye–Waller by virtue of the formal analogy with the eponymous theory of thermal effect in X-ray crystallography, the effect of the transition layer can be made clear: reduction of the ideal reflection coefficient $r_{2,1}$ by a coefficient $exp(-\,2\, e^2\, k_{\perp 1} k_{\perp 2})$ and an increase in the transmission coefficient $t_{2,1}$ by a coefficient $exp\left(\frac{e^2}{2}(k_{\perp 2} - k_{\perp 1})^2\right)$. Interfacial imperfections tend to reduce the reflectance and increase transmittance. They also influence the width of the stopband or in other terms the width of the Bragg peak. This phenomenon has been studied in various works (André and Jonnard 2013).

Figure 5.6 shows the problem: it presents the width of the bandwidth (relative in terms of frequency ω) in *TE* polarization of different orders, in a ratio to the width of the first for a bilayer stratified structure with a period d = 1 and division ratio $\gamma = 1/3$, as a function of the parameter $\delta = \frac{\sigma}{D}$. The quantity $2\sqrt{2}\,\sigma$ quantifies the width 2e of the transition layer (Figure 5.5). The calculation is performed in the context of diffraction theory with two waves, and the values represented by stars appearing in the figure are obtained in the context of a rigorous theory (transfer matrices).

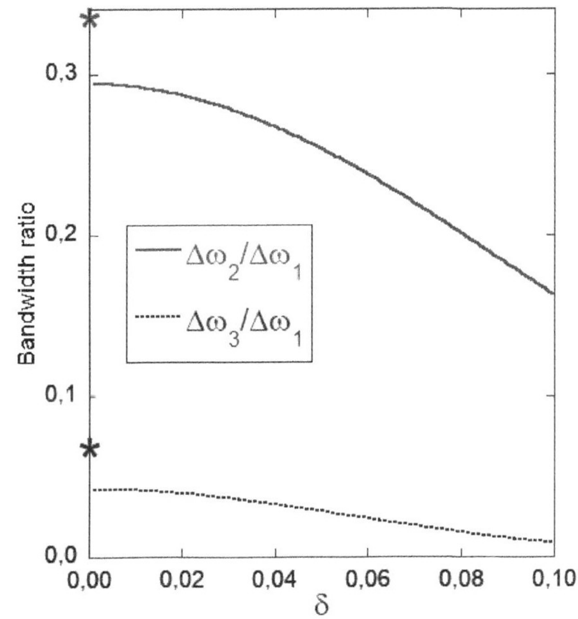

Figure 5.6. *Ratios of the width of the bandwidth (in terms of frequency) $\frac{\Delta\omega_i}{\Delta\omega_1}$ of the second order (curve and red points) and third order (curve and blue points) as a function of the parameter δ for the conditions indicated in the text. For a color version of this figure, see www.iste.co.uk/andre/xrayradiation.zip*

COMMENT ON FIGURE 5.6.– *The starts correspond to the values obtained by a rigorous theory. Adapted with permission from André and Jonnard (2013)*[3]. *© The Optical Society.*

We finish off this section by showing that the Tagaki–Taupin approach (T-T) is effective in modeling the effects of certain structural defects, such as the totally correlated "pure" geometric roughness and the "pure" diffuse interfaces. At this stage and prior to modeling using the T-T approach, we will specify and give details of the concepts of geometric roughness and diffuse interfaces.

Let us consider a periodically stratified medium formed of bilayers; each bilayer is made up of two stacked layers, one formed from medium 1 and the other from medium 2. For a "*pure*" geometric roughness, there is no mix of materials 1 and 2,

3. See doi: 10.1364/JOSAB.30.002296.

but the interface is no longer smooth (flat) but instead forms a "surface" $\Sigma(z, \boldsymbol{r}_{\parallel})$, where $\boldsymbol{r}_{\parallel}$ is a vector of the stratification plane (orthogonal to the stratification axis z).

We can differentiate between the case of total (longitudinal) correlation where *at each interface*, the surfaces $\Sigma(z, \boldsymbol{r}_{\parallel})$ are identical ($\Sigma = \Sigma(\boldsymbol{r}_{\parallel})$); case a) of Figure 5.7 and the case of partial correlation where at each interface, the surface $\Sigma(z, \boldsymbol{r}_{\parallel})$ is different but presents a certain similarity with the other surfaces. This similarity is statistically characterized by a longitudinal correlation function $C_{\perp}$, case b) of Figure 5.6. In the case of *purely* diffuse interfaces, there is horizontal invariance of the geometry of the interfaces, which present smooth "surfaces" Σ which only depend on the variable z ($\Sigma = \Sigma(z)$); case c) of Figure 5.7. In this case, the surface $\Sigma(z)$ is generally described by a profile $\mathrm{Pr}(z)$ given by a slowly varying continuous function. Table 5.2 gives a few examples of functions that generate such profiles. These generating functions are generally functions that tend towards the Heaviside step function when the parameter σ tends to 0. However, we point out that real interfaces are most often rough *and* diffuse, such as those represented in Figure 5.3.

Generating function with profile $\mathrm{Pr}(z)$	Type	Reference
$\frac{1}{2} Erfc(-z/\sigma)$	Névot–Pardo–Corno	Névot et al. (1988)
$1/(1+exp(-z/\sigma))$	Szczyrbowski–Schmalzbauer–Hoffmann	Szczyrbowski et al. (1986)
$\mathrm{Pr}(z) = \left(1/2 + \frac{1}{\pi} tanh(z/\sigma)\right)$	Epstein	Epstein (1930)
$\mathrm{Pr}(z) = \left(1 - exp\left(-\frac{z}{\sigma}\right)\right)^2 \Theta(z)$	Debye–Waller	–

Table 5.2. *Different types of functions used for modeling the profile of a diffuse interface (André 2023)*

The T-T approach to treating pure geometric roughness, or pure interdiffusion, is implemented using the Fourier expansion for dielectric susceptibility; we introduce a generalized development in which the development coefficients become a function of the variable z: $\chi_p \rightarrow \chi_p(z)$ or a phase term $\psi(\epsilon)$ depending on the vertical displacement vector $\boldsymbol{\epsilon}$ of the T-T theory, so in an overall formulation:

$$\chi(z) = \sum_{p=-1}^{1} \chi_p(z)\, exp\big(i\, p\, g + i\, \psi(\epsilon)\big) \qquad [5.34]$$

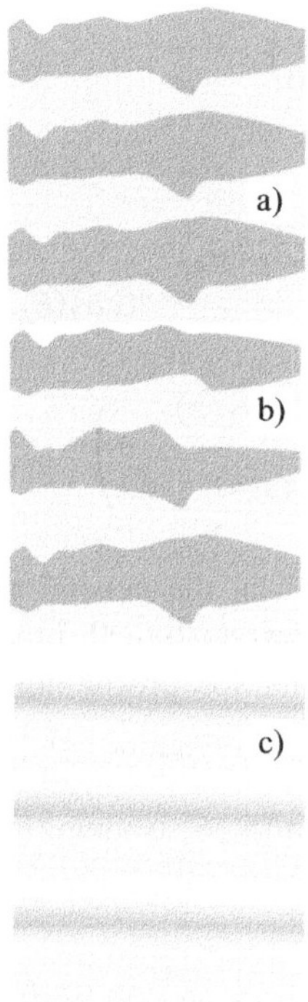

Figure 5.7. *Different interfacial imperfections of a stratified structure formed of bilayers: a) pure roughness with total correlation; b) pure roughness with partial correlation; c) pure interdiffusion (André 2023). For a color version of this figure, see www.iste.co.uk/andre/xrayradiation.zip*

For the case of pure roughness, χ_p remains independent of the variable z, whereas $\psi(\epsilon)$ expresses the phase shift related to the deviation from flatness quantified by ϵ. The vector $\boldsymbol{\epsilon}$ is a function of $\boldsymbol{r}_{\parallel}$ and may depend on the interface considered depending on the nature of the vertical correlation: for total correlation, $\boldsymbol{\epsilon}(\boldsymbol{r}_{\parallel})$ is identical for each interface. For the case of pure interdiffusion, the phase shift $\psi(\epsilon)$ is not involved but χ_p becomes dependent on the variable z. A structure with rough *and* diffuse interfaces can be understood through the overall expression for $\chi(z)$ given by equation [5.34]; the calculation becomes difficult but can be performed using numerical methods (Piault 2015). From a mathematical point of view, the influence of the imperfections on the amplitudes $T(x,z)$ and $R(x,z)$ of the T-T model (see Chapter 3) consists of resolving the system of coupled equations which in matrix form is written with the usual notation (by convenience, $\boldsymbol{r}_{\parallel}$ is denoted as x):

$$\left[\widehat{D} - \widehat{M_{TT}}\right] \begin{pmatrix} T(x,z) \\ R(x,z) \end{pmatrix} = \begin{pmatrix} 0 \\ 0 \end{pmatrix} \qquad [5.35]$$

with:

$$\widehat{D} = \begin{pmatrix} (cos(\theta_B)\,\partial_x + sin(\theta_B)\,\partial_z) & 0 \\ 0 & (cos(\theta_B)\,\partial_x - sin(\theta_B)\,\partial_z) \end{pmatrix}$$

$$\widehat{M_{TT}} = \begin{pmatrix} i\,\frac{k}{2}\,\chi_0(z) & i\,\frac{k}{2}\,\chi_{\bar{1}}(z) \\ i\,\frac{k}{2}\,\chi_1(z) & i\,\frac{k}{2}\,(\chi_0(z) - \tilde{\alpha}) \end{pmatrix} \qquad [5.36]$$

This system is tricky to solve in the general case of rough and interdiffused interfaces but can be useful for analytical processing for the case of pure interdiffusion (where $\partial_x \equiv 0$) and for profile functions such as those listed in Table 5.2.

For the case $\partial_x = 0$, equation [5.36] is simplified and, under Bragg conditions where $\tilde{\alpha} \approx 0$, becomes:

$$\frac{d}{dz}\begin{pmatrix} T(z) \\ R(z) \end{pmatrix} = i\,\widehat{M_{TTz}}\begin{pmatrix} T(z) \\ R(z) \end{pmatrix} \; ; \; \widehat{M_{TTz}} \equiv \begin{pmatrix} \frac{k^2}{2\,k_z}\,\chi_0(z) & \frac{k^2}{2\,k_z}\,\chi_{\bar{1}}(z) \\ \frac{k^2}{2\,k_z}\,\chi_1(z) & -\frac{k^2}{2\,k_z}\,\chi_0(z) \end{pmatrix} \qquad [5.37]$$

$T(z)$ and $R(z)$ have the form:

$$T(z) = C_1(z)\,exp(i\,\widetilde{\mu_1}\,z) + C_2(z)\,exp(i\,\widetilde{\mu_2}\,z)$$

$$R(z) = C_1(z)\,\rho_1\,exp(i\,\widetilde{\mu_1}\,z) + C_2(z)\,\rho_2\,exp(i\,\widetilde{\mu_2}\,z) \qquad [5.38]$$

similar to relationship [3.38] but with $\widetilde{\mu_{1,2}} = \frac{\mu_{1,2}}{sin(\theta_B)}$ and the quantities C_1 and C_2 which depend on the variable z and which satisfy the differential equation:

$$\frac{d}{dz}\begin{pmatrix} C_1(z) \\ C_2(z) \end{pmatrix} = i\,\widetilde{\mathcal{M}_{TTz}}(z)\begin{pmatrix} C_1(z) \\ C_2(z) \end{pmatrix} \qquad [5.39]$$

where:

$$\widehat{\mathcal{M}_{TTz}} = \widehat{\mathcal{P}}^{-1}\,\widehat{\mathcal{T}}^{-1}\,\widehat{\Gamma}\,\widehat{\mathcal{T}}\,\widehat{\mathcal{P}} \qquad [5.40]$$

with:

$$\widehat{\mathcal{P}}(z) = \begin{pmatrix} exp\left(i\,\widetilde{\mu_1}\,z\right) & 0 \\ 0 & exp\left(i\,\widetilde{\mu_2}\,z\right) \end{pmatrix}$$

$$\widehat{\mathcal{T}} = \begin{pmatrix} 1 & 1 \\ \rho_1 & \rho_2 \end{pmatrix}, \widehat{\Gamma} = \begin{pmatrix} \frac{k^2}{2\,k_z} & 0 \\ 0 & -\frac{k^2}{2\,k_z} \end{pmatrix}, \widehat{\Delta} = \begin{pmatrix} \Delta\chi_0(z) & \Delta\chi_{\bar{1}}(z) \\ \Delta\chi_1(z) & \Delta\chi_0(z) \end{pmatrix} \qquad [5.41]$$

The interface profiles in terms of susceptibility are written in the form:

$$\chi_i(z) = \overline{\chi_i} + \Delta\chi_i(z);\ \Delta\chi_i(z) = \Delta\chi_i\,\mathrm{Pr}(z) \qquad [5.42]$$

$\mathrm{Pr}(z)$ is the profile function, as listed in Table 5.2. As a first approximation (order 1 of the expansion in $\Delta\chi_i$), integration of equation [5.39] gives:

$$\begin{pmatrix} C_1(z) \\ C_2(z) \end{pmatrix} = (\mathbb{I} + i\int_0^z \widehat{\mathcal{M}_{TTz}}(Z)\,dZ) \begin{pmatrix} C_1 \\ C_2 \end{pmatrix}_0 \qquad [5.43]$$

where $\begin{pmatrix} C_1 \\ C_2 \end{pmatrix}_0$ corresponds to the case of an abrupt interface. Depending on the profile function $\mathrm{Pr}(z)$, integration can be performed analytically such as in the Debye–Waller case, or numerically. Using this approach, the reflectance can be calculated as well as the angular (excluding specular) distribution, in the presence of a transition layer.

5.2.3. *Practical optimization techniques*

There are few systematic methods which allow the geometric roughness represented in Figure 5.7(a) and (b) to be reduced. One basic rule must be adhered to: using a substrate that is as smooth as possible to avoid reproducing the imperfections in the layers deposited on it by a "memory" effect. Generally, correlation effects at depth are attenuated in layers closer to the surface, if they are sufficient in number. Currently, smooth substrates at atomic (sub-nanometric) scale are available. Electric polarization of the substrate can play a significant role (Zubarev et al. 2008), as can its temperature (Voorma et al. 1998). The technique of ionic polishing is sometimes used to smooth the surfaces generally by bombardment with Ar^+ or Kr^+ ions (Louis et al. 1995). Control of the temperature combined with ionic bombardment has turned out to be positive for improvement of Mo/Si deposits for use in lithography (EUV) (Louis et al. 1994).

The deposition technique (evaporation by ion beams, spraying, etc.) is also an important parameter. Therefore, the pressure of argon in the sputtering technique has a significant influence, and increasing it causes an increase in roughness (Wang et al. 2012; Yi et al. 2017).

Interdiffusion resulting in interfaces such as those represented in Figure 5.7 is combatted using various techniques mainly based on physical, chemical and metallurgical effects. The so-called diffusion barrier technique is highly developed: a thin layer of a suitable material is inserted at the interface between the materials. The boron carbide B_4C and carbon C were used successfully for many bilayer systems. The case Mo/B_4C/Si is well-documented (Maury et al. 2006), in particular in the context of EUV lithography (Braun et al. 2002a, 2002b). A zirconium barrier layer is effective for the Mg/Co system, significant in the EUV domain (Le Guen et al. 2011).

Nitrogenation is another technique often used in the context of reactive sputtering and has turned out to be highly effective in Cr/Sc (Ghafoor et al. 2008) or La/B pairs (Kuznetsov et al. 2015). Modification of the composition of the materials making up the layers also appears to be a promising technique: thus, the addition of rubidium in the silicon layer by forming rubidium silicide would offer thermodynamic stability and better reflectance than pure silicon (Saedi et al. 2020).

Another problem that can affect the performances of a multilayer mirror is mechanical, in other words the stress that may be at the origin of deformation and curvature of the substrate and deposit. This problem is particularly marked when the substrate is an Si_3N_4 membrane used for beamsplitters. To reduce this stress, control of the deposition conditions, argon pressure, temperature, power, etc., is important. Heating after deposition to release the constraints is sometimes effective, in particular for the Mo/Si pair. A detailed review of these questions is available in the article by Lider (2019).

Other sometimes significant characteristics to optimize are the stability of performances over time and the resistance to damage under intense radiation. From the microscopic point of view, the matter constituting the layers acquires energy via the process of photo-absorption: photoelectrons and Auger electrons are emitted which ionize the atoms, and a cascade emission of electrons then occurs (Lipp et al. 2017). We can differentiate between two regimes: non-thermal transition and thermal transition.

In the first, with a timescale of the order of a few hundred femtoseconds, a disturbance of the interatomic potential occurs which can lead to modifications within the materials such as displacement of atoms. Nevertheless, no energy exchange occurs between the hot electrons and the atomic network. In the thermal transition regime, which has a timescale longer than the non-thermal transition timescale, heating of the crystalline lattice due to the energy exchange between the electrons and this lattice occurs, sometimes causing significant structural modifications (fusion). The metals are more susceptible than dielectrics or semi-conductors to this type of transition. Note that both regimes may coexist and interfere (Medvedev et al. 2015).

Concerning the problem of stability, changes in the structure of the multilayers under thermal loading are well documented. Phenomena involving expansion, contraction of layers and interdiffusion are often observed with a correlated change in the reflectance. Almost all the mirrors have been investigated in this way. The reference Oberta et al. (2012) gives an example of the results obtained for mirrors suitable for the hard X-ray radiation (range 6 to 20 keV). The problem of stability primarily arises for imaging systems in the context of industry, in EUV lithography in particular, and for "hostile" or difficult-to-access environments, space telescopes for example. For the case of EUV lithography, particularly important from an economic point of view, oxidation and carbonization of the surface under radiation are a major problem already encountered in optics intended for synchrotron radiation. Klebanoff et al. (2002) proposed to modify the atmosphere inside the building by adding ethanol and water vapor.

Resistance to damage was first studied in the 1990s when it became apparent that multilayer mirrors could be used as constitutive elements of resonant cavities for X-ray lasers produced from power laser-induced plasmas (Ceglio et al. 1986; MacGowan et al. 1992; Barkusky et al. 2010) or in the context of intense sources of X-ray radiation provided by the explosion of nuclear devices (Le Guern et al. 1996, 1997). This involved determining the effects on the structure of stacking under short and unique illumination (pulses lasting for femtoseconds to tens of picoseconds emitted in *mono-shot* mode) and damage thresholds.

Expansion or, on the contrary, contraction phenomena for the periods with profiles varying with depth have been observed, depending on the nature of the materials used, before irreversible damage occurs (Le Guern et al. 1997). Thermodynamic and metallurgical effects and modifications of roughness related to thermal effects have been invoked to explain these phenomena. Damage thresholds are located for fluences of the order of a few tens of mJ/cm^2 and vary considerably according to the type of materials, the X-ray radiation energy and the irradiation geometry (grazing or normal incidence). So-called hardening techniques have been developed, such as the choice of suitable materials (with refractory properties, not very miscible, etc.) or deposition of a protective cap layer made of boron carbide or palladium carbide to try to improve the mirror strength. Slight modification of the thicknesses of surface or deep layers has also been proposed in order to reduce the intensity of the electric field presumed to alter the structure (Hunter and Seely 1993).

Studies on this theme of damage have evolved with the appearance of free electron lasers (FELs) in the X-ray radiation domain. A large number of studies have been completed (Soufli et al. 2011; Kim et al. 2015; Follath et al. 2019) under the conditions of this type of source. For a SASE-FEL (see Chapter 1), there is a periodic train of pulses with an envelope of a total width at half-maximum lasting in the order of 100 fs, sub-divided into sub-fs peaks with an averaged power per pulse in the order of several GW. Different types of unstructured materials in a multilayer stack have been studied: diamond (Gaudin et al. 2013), amorphous carbon (Chalupský et al. 2009), platinum (Krzywinski et al. 2015), inorganic solids (Hau-Riege et al. 2007), etc. Studies have also been performed on multilayer mirrors, in particular on Mo/Si mirrors optimized for EUV lithography (Louis et al. 2009) in which a damage threshold of 45 mJ/cm^2 was measured, and on Pt/C mirrors (Kim et al. 2015) adapted to XFEL in the hard X-ray radiation domain in which a damage threshold of 0.051 mJ/cm^2 has been measured.

5.3. Design of mirrors and inverse problem

5.3.1. *Design of "wide-band" mirrors and super-mirrors*

For mirrors used in imaging tasks, the problem of optimization presents itself above all in terms of integrated reflectance adapted to the angular divergence and to the spectrum of the radiation source. This question appears to be very important in both theoretical and economic terms with the development of EUV lithography, which is considered in Chapter 7. However, conventional EUV Bragg mirrors are not suitable for constituting effective reflectors for this application because their bandwidth, both spectral and angular, is too narrow. For certain applications in astrophysics or metrology, among others, it is also necessary to increase the bandwidths.

The solution generally involves an increase in the angular acceptance of the mirror and of its spectral bandwidth, which requires abandoning the strict periodicity of the structure and searching for a depth profile, in terms of composition and thickness, of a stack of layers offering a wide bandwidth. The problem corresponds to the problem of so-called super-mirrors which will be used to extend the quasi-total reflection zone in the case of grazing incidence for hard X-ray radiation. These super-mirrors are important optics for spatial telescopes in the hard X-ray radiation domain and for light lines from third-generation synchrotron sources delivering high-energy X-ray radiation. "Wide-band" mirrors and super-mirrors are no longer strictly periodic with Bragg-type selective reflection, but instead have an aperiodic profile which, however, is not far removed from the Bragg periodic structure attributed to the spectral domain being reflected. One of the basic ideas consists of performing a quasi-periodic distribution which reflects the softest radiation at the surface of the mirror and the hardest at depth, reducing the quasi-period with depth, in accordance with the scheme presented in Figure 5.8(a).

The technique of a period gradient with depth, designed initially for neutron optics (Mezei 1989), has been used to contribute to astrophysics (Joensen et al. 1993) and the synchrotron radiation (Høghøj et al. 1997; Morawe et al. 2002). Another technique, known as the block method, was proposed by Yamashita et al. (Yamashita et al. 1999; Yao et al. 2013). The basic idea of this type of structure is to ensure a wide spectral band by associating "blocks" of bilayers with different thicknesses and not a period gradient.

In fact, applications of these "wide-band" mirrors and super-mirrors require not only a wide bandwidth but also a "smooth" reflectance in the bandwidth. To satisfy these two criteria, methods based primarily on highly elaborate optimization algorithms have been developed. These algorithms are mainly of two types, local and global. The first correspond to the case where optimal values can be estimated as belonging to a domain known in advance: in this category, we find the Levenberg–Marquardt algorithm (Levenberg 1944; Marquardt 1963), the quasi-Newton method (Chen et al. 2006), the rapid descent method (Petrova and Solov'ev 1997), etc. To improve the convergence of these algorithms, various techniques have been developed to optimize the search for initial values. We select the methods owed to I. Kozhevnikov, which we develop below, and the needle optimization method (Tikhonravov et al. 2007).

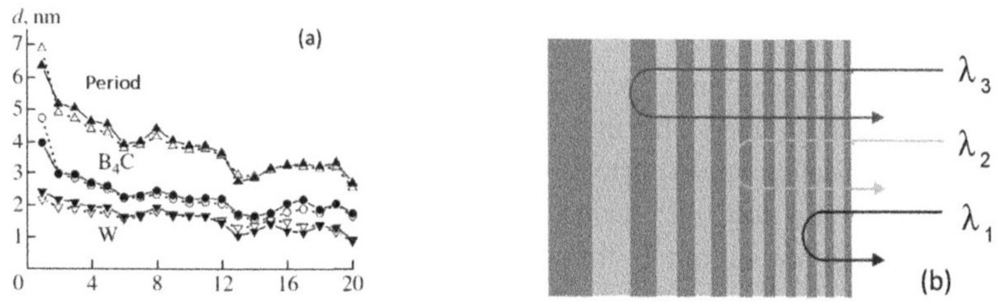

Figure 5.8. *Super-mirror with a period gradient. For a color version of this figure, see www.iste.co.uk/andre/xrayradiation.zip*

COMMENT ON FIGURE 5.8.– *(a) Distribution as a function of the layer number of thicknesses of the layers and of the quasi-period of a super-mirror with period gradient W/B4C, designed to obtain a constant reflectance over a wide angular range at 0.154 nm. Numbering of the bilayers begins at the surface. Adapted from Kozhevnikov et al. (2006)*[4]*, with the permission of Pleiades Publishing, Inc.; (b) chirped mirror, according to Dombi et al. (2005)*[5]*. © The Optical Society*

Global algorithms explore a wider field of values than local algorithms and are generally slower but more powerful in the sense that they can allow values missed by "local algorithms" to be reached. Among the "global algorithms", we can cite simulated annealing and topographical optimization (Dekkers and Aarts 1991; Cheng et al. 2006) with "simplex" algorithms (Nelder and Mead 1965).

In the hard X-ray radiation domain, however, these approaches lead to mirror architectures that are often difficult to create, in particular because they require deposition of ultrafine layers for which interdiffusion problems arise and also because they cause reproducibility problems: hundreds of identical reflector "hubs" are required to form the reflecting module of a space telescope. Using the block method, these problems can be avoided.

In the EUV domain, the problem has also been studied extensively (Pirozhkov and Ragozin 2015), and recently genetic "global algorithms" (Martin et al. 1995) have been developed for this spectral domain (Kuang et al. 2018). Kozhevnikov, working with other researchers (C. Montcalm, I. Bukreeva, E. Ziegler, etc.), made a significant contribution to this problem by combining analytical and numerical approaches. The work concerned has been published in various articles

4. See doi: 10.1134/S1063774506060186.
5. See doi: 10.1364/OPEX.13.010888.

(Kozhevnikov et al. 2001; Kozhevnikov and Montcalm 2010) that we summarize and revise below.

Distribution of the dielectric constant, for an aperiodic structure with an abrupt interface and thickness L, is written using a function $g(z)$ in the following form, with the usual notations for the publication:

$$\varepsilon(z) = \varepsilon_2 + \Delta\varepsilon\, Rect[g(z), \gamma] \qquad [5.44]$$

The function $g(z)$ is increasing and monotonic. It is expressed using the differential function $q(z)$ by:

$$g(z) = \int_0^z q(z')\, dz' \qquad [5.45]$$

The derivative $\frac{dg(z)}{dz}$ is monotonous: if $\frac{dg(z)}{dz} > 0$, the curve is convex and describes a super-mirror; if $\frac{dg(z)}{dz} < 0$, the curve is concave and describes a chirped mirror.

The function $Rect[g(z), \gamma]$ is a periodic rectangle function:

$$Rect[g(z), \gamma] = rect\left[\frac{z}{\gamma\, d_j}\right] ;\, j \text{ index of layer } \in \left[1, \frac{L}{\sum d_j}\right] \qquad [5.46]$$

Taking into account the periodicity of the profile of $\varepsilon(z)$ as a function of $Rect[g(z), \gamma]$, we can write $\varepsilon(z)$ as a Fourier series:

$$\varepsilon(z) = \bar{\varepsilon} + \sum_{n=-\infty}^{+\infty} \varepsilon_n\, exp[i(2\pi\, n\, g(z) + \pi\, n\, \gamma)] \,;\, n \in \mathbb{Z}^* \qquad [5.47]$$

The electric field $E(z)$ is expressed as the superposition of a transmitted and retrograde wave in the same way as in the coupled-wave theory:

$$E(z) = \tilde{\mathcal{F}}(z) exp(i\, \bar{\kappa}\, z) + \tilde{\mathcal{R}}(z)\, exp(-i\, \bar{\kappa}\, z) \qquad [5.48]$$

with $\bar{\kappa} = k\, \sqrt{\bar{\varepsilon} - cos^2(\theta)}$.

Assuming that we can neglect the reflected waves within the structure, which is a reasonable hypothesis outside the zone of total reflection, then using the Green method to resolve the Helmholtz equation, we obtain:

$$\tilde{\mathcal{F}}(z) = 1 + \frac{i\, k^2}{2\, \bar{\kappa}} \int_0^Z E(z')(\varepsilon(z') - \bar{\varepsilon})\, exp(-i\, \bar{\kappa}\, z')\, dz'$$

$$\tilde{\mathcal{R}}(z) = 1 + \frac{i\, k^2}{2\, \bar{\kappa}} \int_Z^L E(z')(\varepsilon(z') - \bar{\varepsilon})\, exp(i\, \bar{\kappa}\, z')\, dz' \qquad [5.49]$$

Combining equations [5.45] to [5.49], we obtain a system of coupled complete equations:

$$\tilde{\mathcal{F}}(z) = 1 + \frac{i\,k^2}{4\,\bar{\kappa}} \sum_{n=-\infty}^{+\infty} \varepsilon_n\, exp[i\,\pi\,n\,\gamma] \int_0^Z \tilde{\mathcal{R}}(z')(\varepsilon(z') - \bar{\varepsilon})\, exp(i\,2\pi\,n\,g(z') - i\,2\,\bar{\kappa}\,z')\,dz' + S_{\tilde{\mathcal{F}}}(z)$$

$$\tilde{\mathcal{R}}(z) = 1 + \frac{i\,k^2}{4\,\bar{\kappa}} \sum_{n=-\infty}^{+\infty} \varepsilon_n\, exp[-i\,\pi\,n\,\gamma] \int_0^Z \tilde{\mathcal{F}}(z')(\varepsilon(z') - \bar{\varepsilon})\, exp(-i\,2\pi\,n\,g(z') + i\,2\,\bar{\kappa}\,z')\,dz' + S_{\tilde{\mathcal{R}}}(z) \quad [5.50]$$

The terms $S_{\tilde{\mathcal{F}}}(z)$ and $S_{\tilde{\mathcal{R}}}(z)$ group together all the terms that are not explicitly developed in equation [5.50]. In super-mirror or chirped mirror structures with a period gradient (see Figure 5.8(b)), Bragg resonance only occurs for a restricted part of the bilayers which, for a given radiation energy, meet the Bragg condition. Outside this portion, the structure behaves as an attenuator of the incident radiation. From a mathematical point of view, this means that the resonance is described by rapidly oscillating terms of the integral elements in equation [5.50], such that the value of the integrals is given by stationary points[6] and the terms $S_{\tilde{\mathcal{F}}}(z)$ and $S_{\tilde{\mathcal{R}}}(z)$ do not contribute. Employing the stationary phase method, a complex calculation is used to estimate the integral $I(z) = \int_0^Z \tilde{\mathcal{F}}(z')(\varepsilon(z') - \bar{\varepsilon})\ exp(-i\,2\pi\,n\,g(z') + i\,2\,\bar{\kappa}\,z')\,dz$ involved in the expression of $\tilde{\mathcal{R}}(z)$ in equation [5.49] as follows:

$$I(z) \approx \frac{\tilde{\mathcal{F}}(z)}{\sqrt{2|q'(z)|}}\, exp(-i\,2\pi\,n\,g(z) + i\,2\,\bar{\kappa}\,z)\, G\left(z\,\sqrt{2|q'(z)|}\right) \quad [5.51]$$

where:

$$G(x) = C(x) - i\,sign(q'(z))\,S(x) \quad [5.52]$$

$C(x)$ and $S(x)$ are Fresnel integrals (see Chapter 4). The other integral can be evaluated in the same way. Introducing the approximate expressions for the integrals into system [5.50], we obtain the following system of coupled approximate equations:

$$\tilde{\mathcal{F}}(z) = 1 + \frac{i\,k^2}{4\,\bar{\kappa}}\,\varepsilon_1\,\frac{\tilde{\mathcal{R}}(z)}{\sqrt{2|q'(z)|}}\,G^*\left(z\,\sqrt{2|q'(z)|}\right) exp(+i\,2\pi\,n\,g(z) - i\,2\,\bar{\kappa}\,z + i\,\pi\,\gamma)$$

$$\tilde{\mathcal{R}}(z) = \frac{i\,k^2}{4\,\bar{\kappa}}\,\varepsilon_1\,\frac{\tilde{\mathcal{F}}(z)}{\sqrt{2|q'(z)|}}\,G\left((L - z)\,\sqrt{2|q'(z)|}\right) exp(-i\,2\pi\,n\,g(z) i\,2\,\bar{\kappa}\,z - i\,\pi\,\gamma) \quad [5.53]$$

6. The stationary phase method is a tool commonly used to integrate oscillating functions.

Solving system [5.53] and moving to the limit of the Fresnel integrals[7], the reflection coefficient $r \approx \tilde{\mathcal{R}}(0)$ can be expressed, and an expression approaching the reflectance R as a function of the wavelength λ is given:

$$R(\lambda) \approx \left|\tilde{\mathcal{R}}(0)\right|^2 \approx \left|\frac{2\,\eta\,\sqrt{2|q'(z)|}}{\eta^2+|q'(z)|}\right|^2 exp\big(-\,i\,4\,Im(\bar{\kappa})\,z(\lambda)\big);\ \eta = \frac{k^2}{8\,\bar{\kappa}}\,\varepsilon_1 \quad [5.54]$$

It is important to highlight that $z(\lambda)$ is not an independent variable but instead a stationary point satisfying:

$$\frac{dg(z)}{dz} = \frac{Re(\bar{\kappa})}{n\,\pi} \quad [5.55]$$

Relationship [5.54] describes the spectral or angular dependence of the reflectance of a structure with period gradient with depth. Its validity presumes that the reflection takes place outside the zone of total reflection and that the gradient of the thickness distribution is large enough. It is intended to be used as a basic expression in the context of resolution of an inverse problem. In this case, this problem consists of finding the function $g(z)$ which is adjusted to a chosen, predetermined variation of reflectance in terms of angle or wavelength. The strategy is as follows: equation [5.54] is transformed into a second-degree equation in $\frac{dq(z)}{dz} \equiv q'(z)$, for which the selected physical solution is:

$$|q'| = \eta^2\,\frac{2-\tau+\sqrt{1-\tau}}{\tau} \equiv F(\lambda,z);\ \tau \equiv R(\lambda)\,exp\big(\,i\,4\,Im(\bar{\kappa})\,z(\lambda)\big) \quad [5.56]$$

and which is valid primarily in the domain of hard X-ray radiation because $Im(\eta)$ has been neglected. From equation [5.56] and the Bragg law, we deduce the relationship:

$$\frac{d\lambda}{dz} = \pm\pi\,F(\lambda,z)\,\left|\frac{dRe(\bar{\kappa})}{d\lambda}\right|^{-1} \quad [5.57]$$

It constitutes a correlation between the variation with depth of the diffracted wavelengths and the term $F(\lambda, z)$. This relationship can be used as an analytical starting point for solving the inverse problem with a numerical method, using one of the methods or algorithms previously presented.

7. The asymptotic limit value of (non-normalized) Fresnel integrals $C(\infty)$ and $S(\infty)$ is equal to $\frac{1}{2}\sqrt{\frac{\pi}{2}}$.

5.3.2. *Design of chirped mirrors*

The first condition to fulfill is conservation of the complete spectrum of the ultrashort pulse, which can extend across several dozen eV, which requires wide-band reflection. As in the case of super-mirrors, we use an aperiodic structure with period gradient (Ménesguen et al. 2010) where the algorithms analogous to those used for super-mirrors can be involved in the design. However, aside from the problem of optimizing the width of the spectral band, the additional problem of controlling the phase arises. We have the choice between:

– reflecting the pulse without deforming it, which implies that the mirror must have a linear spectral phase; this question was tackled by Morlens et al. (2006);

– compensating for the *attochip*[8] with resynchronization of harmonics, which can be done with a convex parabolic spectral phase (Wonisch et al. 2004; Morlens et al. 2005).

5.4. References

Alehyane, N., Arbaoui, M., Barchewitz, R., André, J.-M., Christensen, F.E., Hornstrup, A., Palmari, J., Rasigni, M., Rivoira, R., Rasigni, G. (1989). Extreme UV and X-ray scattering measurements from a rough LiF crystal surface characterized by electron micrography. *Applied Optics*, 28(10), 1763–1772.

André, J.-M. (1984). Rugosité de surface et réflexion spéculaire du rayonnement X-UV. *Optics Communications*, 52(2), 87–93.

André, J.-M. (2023). *Cristaux et quasi-cristaux photoniques unidimensionnels : principe et théorie*. Auto-édition JMA, Amboise.

André, J.-M. and Barchewitz, R. (2002). X-ray multilayer optics: A review. In *Recent Research Devolpments in Optics*, Pandalai, S.G. (ed.). Research Signpost, Kerala, 645–643.

André, J.-M. and Jonnard, P. (2013). Photonic bandgaps of periodic multilayers with diffuse interfaces. *Journal of the Optical Society of America B*, 30(8), 2296–2300.

8. The attochip is a phenomenon resulting from the genesis of ultrashort pulses (attoseconds) in the X-UV domain. These pulses are generated by a nonlinear optical process known as high-harmonic generation (Ruchon et al. 2010); lower-order harmonics are emitted earlier than those of higher order, which causes desynchronization in the emission of harmonics, giving rise to the attochip.

André, J.-M., Benbalagh, R., Barchewitz, R., Ravet, M.-F., Raynal, A., Delmotte, F., Bridou, F., Julié, G., Bosseboeuf, A., Laval, R. et al. (2002). X-ray multilayer monochromator with enhanced performance. *Applied Optics*, 41(1), 239–244.

Barkusky, F., Bayer, A., Döring, S., Grossmann, P., Mann, K. (2010). Damage threshold measurements on EUV optics using focused radiation from a table-top laser produced plasma source. *Optics Express*, 18(5), 4346–4355.

Barysheva, M.M., Garakhin, S.A., Zuev, S.Y., Polkovnikov, V.N., Salashchenko, N.N., Svechnikov, M.V., Chkhalo, N.I., Yulin, S. (2019). Comparison of approaches in the manufacture of broadband mirrors for the EUV range: Aperiodic and stack structures. *Quantum Electronics*, 49(4), 380.

Benbalagh, R., André, J.-M., Barchewitz, R., Jonnard, P., Julié, G., Mollard, L., Rolland, G., Rémond, C., Troussel, P., Marmoret, R. et al. (2005). Lamellar multilayer amplitude grating as soft-X-ray Bragg monochromator. *Nuclear Instruments and Methods in Physics Research Section A: Accelerators, Spectrometers, Detectors and Associated Equipment*, 541(3), 590–597.

Bennett, J.M. (1976). Measurement of the rms roughness, autocovariance function and other statistical properties of optical surfaces using a FECO scanning interferometer. *Applied Optics*, 15(11), 2705–2721.

de Boer, D.K.G. (1995). X-ray reflection and transmission by rough surfaces. *Physical Review B*, 51(8), 5297–5305.

Boher, P., Hennet, L., Houdy, Ph. (1990). Three materials soft X-ray mirrors – Theory and application. *Proceedings of SPIE*, 1345, 198–212.

Bourassin-Bouchet, C., Diveki, Z., de Rossi, S., English, E., Meltchakov, E., Gobert, O., Guenot, D., Carre, B., Delmotte, F., Ruchon, T. et al. (2011). Optiques pour les impulsions attosecondes. In *UVX 2010 – 10ème Colloque sur les Sources Cohérentes et Incohérentes UV, VUV et X ; Applications et Développements Récents*. EDP Sciences, Ile de Porquerolles, 35–40 [Online]. Available at: http://uvx.edpsciences.org/10.1051/uvx/2011005 [Accessed 20 January 2023].

Braun, S., Mai, H., Moss, M., Scholz, R. (2002a). Microstructure of Mo/Si multilayers with barrier layers. *Proceedings of SPIE*, 4782, 185.

Braun, S., Mai, H., Moss, M., Scholz, R., Leson, A. (2002b). Mo/Si multilayers with different barrier layers for applications as extreme ultraviolet mirrors. *Japanese Journal of Applied Physics*, 41(6S), 4074.

Carniglia, C.K. and Apfel, J.H. (1980). Maximum reflectance of multilayer dielectric mirrors in the presence of slight absorption. *JOSA*, 70(5), 523–534.

Ceglio, N.M., Stearns, D.G., Hawryluk, A.M., Barbee, T.W., Danzmann, K., Kuhne, M., Mueller, P., Wende, B., Stearns, M.B., Petford-Long, A.K. et al. (1986). Soft X-ray laser cavities. *Le Journal de Physique Colloques*, 47(C6), C6-277–C6-286.

Chalupský, J., Hájková, V., Altapova, V., Burian, T., Gleeson, A.J., Juha, L., Jurek, M., Sinn, H., Störmer, M., Sobierajski, R. et al. (2009). Damage of amorphous carbon induced by soft X-ray femtosecond pulses above and below the critical angle. *Applied Physics Letters*, 95(3), 031111.

Chen, L.H., Deng, N.Y., Zhang, J.Z. (2006). A modified quasi-Newton method for structured optimization with partial information on the Hessian. *Computational Optimization and Applications*, 35(1), 5–18.

Cheng, X., Wang, Z., Zhang, Z., Wang, F., Chen, L. (2006). Design of X-ray super-mirrors using simulated annealing algorithm. *Optics Communications*, 265(1), 197–206.

Daillant, J. and Bélorgey, O. (1992). Surface scattering of X-rays in thin films. Part I. Theoretical treatment. *The Journal of Chemical Physics*, 97(8), 5824–5836.

Dekkers, A. and Aarts, E. (1991). Global optimization and simulated annealing. *Mathematical Programming*, 50(1), 367–393.

Dombi, P., Yakovlev, V.S., O'Keeffe, K., Fuji, T., Lezius, M., Tempea, G. (2005). Pulse compression with time-domain optimized chirped mirrors. *Optics Express*, 13(26), 10888.

Epstein, P.S. (1930). Reflection of waves in an inhomogeneous absorbing medium. *Proceedings of the National Academy of Sciences*, 16(10), 627–637.

Fialin, M., Rémy, H., André, J.M., Chauvineau, J.P., Rousseaux, F., Ravet, M.F., Decanini, D., Cambril, E. (1996). Extending the possibilities of soft X-ray spectrometry through the etching of layered synthetic microstructure monochromators. *X-Ray Spectrometry*, 25(2), 60–65.

Follath, R., Koyama, T., Lipp, V., Medvedev, N., Tono, K., Ohashi, H., Patthey, L., Yabashi, M., Ziaja, B. (2019). X-ray induced damage of B4C-coated bilayer materials under various irradiation conditions. *Scientific Reports*, 9(1), 2029.

Freitag, J.M. and Clemens, B.M. (2000). Nonspecular X-ray reflectivity study of roughness scaling in Si/Mo multilayers. *Journal of Applied Physics*, 89(2), 1101–1107.

Furman, S.A. and Tikhonravov, A.V. (1992). *Basics of Optics of Multilayer Systems*. Frontières, Gif-sur-Yvette.

Gaudin, J., Medvedev, N., Chalupský, J., Burian, T., Dastjani-Farahani, S., Hájková, V., Harmand, M., Jeschke, H.O., Juha, L., Jurek, M. et al. (2013). Photon energy dependence of graphitization threshold for diamond irradiated with an intense XUV FEL pulse. *Physical Review B*, 88(6), 060101.

Ghafoor, N., Persson, P.O.Å., Birch, J., Eriksson, F., Schäfers, F. (2006). Interface engineered ultrashort period Cr–Ti multilayers as high reflectance mirrors and polarizers for soft X-rays of lambda = 2.74 nm wavelength. *Applied Optics*, 45(1), 137–143.

Ghafoor, N., Eriksson, F., Gullikson, E., Hultman, L., Birch, J. (2008). Incorporation of nitrogen in Cr/Sc multilayers giving improved soft X-ray reflectivity. *Applied Physics Letters*, 92(9), 091913.

Gibaud, A. and Hazra, S. (2000). X-ray reflectivity and diffuse scattering. *Current Science*, 78(12), 1467–1477.

Guichet, C., Rivoira, R., Rasigni, G., André, J.-M., Barchewitz, R. (1997). Influence of the number of bilayers on the optical performances of X-UV multilayer interferential mirrors. *Journal of X-Ray Science and Technology*, 7(1), 24–36.

Haase, A., Soltwisch, V., Braun, S., Laubis, C., Scholze, F. (2017). Interface morphology of Mo/Si multilayer systems with varying Mo layer thickness studied by EUV diffuse scattering. *Optics Express*, 25(13), 15441–15455.

Hau-Riege, S.P., London, R.A., Bionta, R.M., McKernan, M.A., Baker, S.L., Krzywinski, J., Sobierajski, R., Nietubyc, R., Pelka, J.B., Jurek, M. et al. (2007). Damage threshold of inorganic solids under free-electron-laser irradiation at 32.5 nm wavelength. *Applied Physics Letters*, 90(17), 173128.

Høghøj, P., Ziegler, E., Susini, J., Freund, A.K., Joensen, K.D., Gorenstein, P., Wood, J.L. (1997). Focusing of hard X-rays with a WSi supermirror. *Nuclear Instruments and Methods in Physics Research Section B: Beam Interactions with Materials and Atoms*, 132(3), 528–533.

Holý, V. and Baumbach, T. (1994). Nonspecular X-ray reflection from rough multilayers. *Physical Review B*, 49(15), 10668–10676.

Hunter, W.R. and Seely, J.F. (1993). Low-radiation-damage multilayer coatings for the soft-X-ray region. *Applied Optics*, 32(25), 4846–4851.

Joensen, K.D., Christensen, F.E., Schnopper, H.W., Gorenstein, P., Susini, J., Hoghoj, P., Hustache, R., Wood, J.L., Parker, K. (1993). Medium-sized grazing incidence high-energy X-ray telescopes employing continuously graded multilayers. *Proceedings of SPIE*, 1736, 239–248.

Joensen, K.D., Hoghoj, P., Christensen, F.E., Gorenstein, P., Susini, J., Ziegler, E., Freund, A.K., Wood, J.L. (1994). Multilayered supermirror structures for hard X-ray synchrotron and astrophysics instrumentation. *Proceedings of SPIE*, 2011, 360–372.

Joensen, K.D., Voutov, P., Szentgyorgyi, A., Roll, J., Gorenstein, P., Hoghoj, P., Christensen, F.E. (1995). Design of grazing-incidence multilayer supermirrors for hard-X-ray reflectors. *Applied Optics*, 34(34), 7935–7944.

Jonnard, P., Le Guen, K., André, J.-M. (2009). High-resolution X-ray analysis with multilayer gratings. *X-Ray Spectrometry*, 38(2), 117–120.

Karpov, S.Y. and Stolyarov, S.N. (1993). Propagation and transformation of electromagnetic waves in one-dimensional periodic structures. *Physics-Uspekhi*, 36(1), 1–22.

Kim, J., Nagahira, A., Koyama, T., Matsuyama, S., Sano, Y., Yabashi, M., Ohashi, H., Ishikawa, T., Yamauchi, K. (2015). Damage threshold of platinum/carbon multilayers under hard X-ray free-electron laser irradiation. *Optics Express*, 23(22), 29032–29037.

Klebanoff, L.E., Clift, W.M., Malinowski, M.E., Steinhaus, C., Grunow, P., Bajt, S. (2002). Radiation-induced protective carbon coating for extreme ultraviolet optics. *Journal of Vacuum Science & Technology B: Microelectronics and Nanometer Structures Processing, Measurement, and Phenomena*, 20(2), 696–703.

Kozhevnikov, I.V. and Montcalm, C. (2010). Design of X-ray multilayer mirrors with maximal integral efficiency. *Nuclear Instruments and Methods in Physics Research Section A: Accelerators, Spectrometers, Detectors and Associated Equipment*, 624(1), 192–202.

Kozhevnikov, I.V. and Vinogradov, A.V. (1987). Basic formulae of XUV multilayer optics. *Physica Scripta*, 1987(T17), 137.

Kozhevnikov, I.V., Bukreeva, N., Ziegler, E. (2001). Design of X-ray supermirrors. *Nuclear Instruments and Methods in Physics Research Section A: Accelerators, Spectrometers, Detectors and Associated Equipment*, 460(2), 424–443.

Kozhevnikov, I.V., Voronov, A.S., Roshchin, B.S., Asadchikov, V.E., Mednikov, K.N., Pirozhkov, A.S., Ragozin, E.N., Wang, Zh., Zhong, Zh., Wang, F. (2006). Design, fabrication, and study of wideband multilayer X-ray mirrors. *Crystallography Reports*, 51(6), 1075–1081.

Kozhevnikov, I.V., van der Meer, R., Bastiaens, H.M.J., Boller, K.-J., Bijkerk, F. (2010). High-resolution, high-reflectivity operation of lamellar multilayer amplitude gratings: Identification of the single-order regime. *Optics Express*, 18(15), 16234–16242.

Krzywinski, J., Cocco, D., Moeller, S., Ratner, D. (2015). Damage threshold of platinum coating used for optics for self-seeding of soft X-ray free electron laser. *Optics Express*, 23(5), 5397–5405.

Kuang, S., Gong, X., Yang, H. (2018). Robust design of broadband EUV multilayer using multi-objective evolutionary algorithm. *Optics Communications*, 410, 805–810.

Kushnir, V.I. and Suvorov, E.V. (1990). X-ray backscattering on perfect crystals ($2\theta \approx \pi$). *Physica Status Solidi (a)*, 122(1), 391–404.

Kuznetsov, D.S., Yakshin, A.E., Sturm, J.M., van de Kruijs, R.W.E., Louis, E., Bijkerk, F. (2015). High-reflectance La/B-based multilayer mirror for 6.x nm wavelength. *Optics Letters*, 40(16), 3778–3781.

Larruquert, J.I. (2001). General theory of sub-quarterwave multilayers with highly absorbing materials. *JOSA A*, 18(10), 2617–2627.

Larruquert, J.I. (2002a). Reflectance enhancement in the extreme ultraviolet and soft X-rays by means of multilayers with more than two materials. *JOSA A*, 19(2), 391–397.

Larruquert, J.I. (2002b). Sub-quarterwave multilayers with enhanced reflectance at 13.4 and 11.3 nm. *Optics Communications*, 206(4), 259–273.

Le Guern, F., Troussel, P., André, J.-M., Friart, D., Jalinaud, T., Rabec le Gloahec, M., Desenne, D., Le Breton, J.-P., Boutin, J.-Y., Dutrannoy, J.-L. et al. (1996). Damages to B4C/W multilayer mirrors by intense soft X-ray bursts. *Review of Scientific Instruments*, 67(6), 2107–2110.

Le Guern, F., André, J.-M., Lebreton, J.-P., Dutrannoy, J.-L., Chauvineau, J.-P., Krastev, K., Larcade, J.-L., Nazet, C., Barchewitz, R. (1997). Experimental study and simulation of the damage induced to various multilayer interferential mirrors by soft X-ray plasma-laser sources. *Journal of X-Ray Science and Technology*, 7(3), 271–283.

Le Guern, F., Troussel, P., André, J.-M., Friart, D., Jalinaud, T., Rabec le Gloahec, M., Desenne, D., Le Breton, J.-P., Boutin, J.-Y., Dutrannoy, J.-L. et al. (2011). Introduction of Zr in nanometric periodic Mg/Co multilayers. *Applied Physics A*, 102(1), 69–77.

Levenberg, K. (1944). A method for the solution of certain non-linear problems in least squares. *Quarterly of Applied Mathematics*, 2(2), 164–168.

Lider, V.V. (2019). Multilayer X-ray interference structures. *Physics-Uspekhi*, 62(11), 1063.

Lipp, V., Medvedev, N., Ziaja, B. (2017). Classical Monte-Carlo simulations of x-ray induced electron cascades in various materials. *Proceedings of SPIE*, 10236, 102360H.

Louis, E., Voorma, H.-J., Koster, N.B., Shmaenok, L., Bijkerk, F., Schlatmann, R., Verhoeven, J., Platonov, Yu.Ya., van Dorssen, G.E., Padmore, H.A. (1994). Enhancement of reflectivity of multilayer mirrors for soft X-ray projection lithography by temperature optimization and ion bombardment. *Microelectronic Engineering*, 23(1), 215–218.

Louis, E., Spiller, E.A., Abdali, S., Christensen, F.E., Voorma, H.-J., Koster, N.B., Frederiksen, P.K., Tarrio, C., Gullikson, E.M., Bijkerk, F. (1995). Multilayer X-ray mirrors for the objective crystal spectrometer on the Spectrum Roentgen Gamma satellite. *Proceedings of SPIE*, 2515, 194.

Louis, E., Khorsand, A.R., Sobierajski, R., van Hattum, E.D., Jurek, M., Klinger, D., Pelka, J.B., Juha, L., Chalupský, J., Cihelka, J. et al. (2009). Damage studies of multilayer optics for XUV free electron lasers. *Proceedings of SPIE*, 7361, 73610I-73610I–6.

MacGowan, B.J., Mrowka, S., Barbee Jr., T.W., da Silva, L.B., Eder, D.C., Koch, J.A., Pan, L.S., Turner, J.A., Underwood, J.H., Young, P.E. (1992). Investigation of damage to multilayer optics in X-ray laser cavities: W/C, WRe/C, WC/C, stainless-steel/C, and Cr3C2/C mirrors. *Journal of X-Ray Science and Technology*, 3(4), 231–282.

Marquardt, D.W. (1963). An algorithm for least-squares estimation of nonlinear parameters. *Journal of the Society for Industrial and Applied Mathematics*, 11(2), 431–441.

Martin, S., Rivory, J., Schoenauer, M. (1995). Synthesis of optical multilayer systems using genetic algorithms. *Applied Optics*, 34(13), 2247–2254.

Martynov, V.V., Platonov, Yu., Kazimirov, A., Bilderback, D.H. (2004). High-resolution X-ray multilayers. *AIP Conference Proceedings*, 705(1), 697–700.

Maury, H., Jonnard, P., André, J.-M., Gautier, J., Roulliay, M., Bridou, F., Delmotte, F., Ravet, M.-F., Jérome, A., Holliger, P. (2006). Non-destructive X-ray study of the interphases in Mo/Si and Mo/B4C/Si/B4C multilayers. *Thin Solid Films*, 514(1/2), 278–286.

Medvedev, N., Li, Z., Ziaja, B. (2015). Thermal and nonthermal melting of silicon under femtosecond X-ray irradiation. *Physical Review B*, 91(5), 054113.

Ménesguen, Y., de Rossi, S., Meltchakov, E., Delmotte, F. (2010). Aperiodic multilayer mirrors for efficient broadband reflection in the extreme ultraviolet. *Applied Physics A – Materials Science & Processing*, 98(2), 305–309.

Mezei, F. (1989). Very high reflectivity supermirrors and their applications. *Proceedings of SPIE*, 983, 10.

Morawe, C., Ziegler, E., Peffen, J.-C., Kozhevnikov, I.V. (2002). Design and fabrication of depth-graded X-ray multilayers. *Nuclear Instruments and Methods in Physics Research Section A: Accelerators, Spectrometers, Detectors and Associated Equipment*, 493(3), 189–198.

Morlens, A.-S., Balcou, P., Zeitoun, P., Valentin, C., Laude, V., Kazamias, S. (2005). Compression of attosecond harmonic pulses by extreme-ultraviolet chirped mirrors. *Optics Letters*, 30(12), 1554–1556.

Morlens, A.-S., López-Martens, R., Boyko, O., Zeitoun, P., Balcou, P., Varjú, K., Gustafsson, E., Remetter, T., L'Huillier, A., Kazamias, S. et al. (2006). Design and characterization of extreme-ultraviolet broadband mirrors for attosecond science. *Optics Letters*, 31(10), 1558–1560.

Nelder, J.A. and Mead, R. (1965). A simplex method for function minimization. *The Computer Journal*, 7(4), 308–313.

Névot, L., Pardo, B., Corno, C. (1988). Analyse des matériaux stratifiés par réflectométrie X. *Analusis*, 16(7), 381–388.

Oberta, P., Platonov, Y., Flechsig, U. (2012). Investigation of multilayer X-ray optics for the 6 keV to 20 keV energy range. *Journal of Synchrotron Radiation*, 19(5), 675–681.

Pardo, B., Megademini, T., André, J.-M. (1988). X-UV synthetic interference mirrors: Theoretical approach. *Revue de Physique Appliquée*, 23(10), 1579–1597.

Petrova, S.S. and Solov'ev, A.D. (1997). The origin of the method of steepest descent. *Historia Mathematica*, 24(4), 361–375.

Piault, P. (2015). Modélisation numérique et caractérisation des défauts dans les miroirs multicouches en vue de leur application en imagerie X cohérente. PhD Thesis, Paris-Saclay University, Paris.

Pirozhkov, A.S. and Ragozin, E.N. (2015). Aperiodic multilayer structures in soft X-ray optics. *Physics-Uspekhi*, 58(11), 1095–1105.

Rasigni, M., Rasigni, G., Palmari, J.-P., Llebaria, A. (1981). Study of surface roughness using microdensitometer analysis of electron micrographs of surface replicas. II: Autocovariance functions. *JOSA*, 71(10), 1230–1237.

Rosenbluth, A.E. (1988). Computer search for layer materials that maximize the reflectivity of X-ray multilayers. *Revue de Physique Appliquée*, 23(10), 1599–1621.

Ruchon, T., Salières, P., Zeitoun, P., Sebban, S. (2010). Sources cohérentes de laboratoire dans l'extrême ultraviolet. *Reflets de la physique*, 30–34.

Saedi, M., Sfiligoj, C., Verhoeven, J., Frenken, J.W.M. (2020). Effect of rubidium incorporation on the optical properties and intermixing in Mo/Si multilayer mirrors for EUV lithography applications. *Applied Surface Science*, 507, 144951.

Salditt, T., Lott, D., Metzger, T.H., Peisl, J., Vignaud, G., Höghöj, P., Schärpf, O., Hinze, P., Lauer, R. (1996). Interfacial roughness and related growth mechanisms in sputtered W/Si multilayers. *Physical Review B*, 54(8), 5860–5872.

Sammar, A., André, J.-M., Ouahabi, M., Pardo, B., Barchewitz, R. (1993). Monochromateur multicouche à bande passante étroite pour le rayonnement X. *Comptes Rendus Académie des Sciences, Série II*, 316, 1055–1060.

Singh, M. and Braat, J.J.M. (2000). Design of multilayer extreme-ultraviolet mirrors for enhanced reflectivity. *Applied Optics*, 39(13), 2189–2197.

Sinha, S.K. (1996). Surface roughness by X-ray and neutron scattering methods. *Acta Physica Polonica A*, 89(2), 219–234.

Sinha, S.K., Sirota, E.B., Garoff, S., Stanley, H.B. (1988). X-ray and neutron scattering from rough surfaces. *Physical Review B*, 38(4), 2297–2311.

Soufli, R., Fernández-Perea, M., Hau-Riege, S.P., Baker, S.L., Robinson, J.C., Gullikson, E.M., Bozek, J.D., Kelez, N.M., Boutet, S. (2011). Lifetime and damage threshold properties of reflective X-ray coatings for the LCLS free-electron laser. *Proceedings of SPIE*, 8077, 807702.

Störmer, M., André, J.M., Michaelsen, C., Benbalagh, R., Jonnard, P. (2007). X-ray scattering from etched and coated multilayer gratings. *Journal of Physics*, 40, 4253–4258. doi: 10.1088/0022-3727/40/14/022.

Szczyrbowski, J., Schmalzbauer, K., Hoffmann, H. (1986). Optical properties of In2O3–SnO2 films in the transparent and plasma reflection region. *Thin Solid Films*, 137(2), 169–183.

Tikhonravov, A.V., Trubetskov, M.K., DeBell, G.W. (2007). Optical coating design approaches based on the needle optimization technique. *Applied Optics*, 46(5), 704–710.

Vinogradov, A.V. and Zeldovich, B.Y. (1977). X-ray and far UV multilayer mirrors: Principles and possibilities. *Applied Optics*, 16(1), 89–93.

Voorma, H.-J., Louis, E., Koster, N.B., Bijkerk, F. (1998). Temperature induced diffusion in Mo/Si multilayer mirrors. *Journal of Applied Physics*, 83(9), 4700–4708.

Wang, F., Zhu, J., Qi, Z., Wang, Z., Jonnard, P., Le Guen, K., André, J.-M., Fialin, M. (2012). Influence of working gas pressure on the performance of W/Si multilayers. *Chinese Physics C*, 36(9), 909.

Wonisch, A., Westerwalbesloh, T., Hachmann, W., Kabachnik, N., Kleineberg, U., Heinzmann, U. (2004). Aperiodic nanometer multilayer systems as optical key components for attosecond electron spectroscopy. *Thin Solid Films*, 464–465, 473–477.

Yamamoto, M. and Namioka, T. (1992). Layer-by-layer design method for soft-X-ray multilayers. *Applied Optics*, 31(10), 1622–1630.

Yamashita, K., Akiyama, K., Haga, K., Kunieda, H., Lodha, G.S., Nakajo, N., Nakamura, N., Okajima, T., Tamura, K., Tawara, Y. (1998). Fabrication and characterization of multilayer supermirrors for hard X-ray optics. *Journal of Synchrotron Radiation*, 5(3), 711–713.

Yamashita, K., Kunieda, H., Tawara, Y., Tamura, K., Ogasaka, Y., Haga, K., Okajima, T., Hidaka, Y., Ichimaru, S., Takahashi, S. et al. (1999). New design concept of multilayer supermirrors for hard X-ray optics. *Proceedings of SPIE*, 3766, 327–335.

Yao, Y., Kunieda, H., Wang, Z. (2013). The theoretical analysis of the hard X-ray block-structure supermirror. *Optics Express*, 21(7), 8638–8651.

Yi, Q., Huang, Q., Wang, X., Yang, Y., Yang, X., Zhang, Z., Wang, Z., Xu, R., Peng, T. et al. (2017). Structure and extreme ultraviolet performance of Si/C multilayers deposited under different working pressures. *Applied Optics*, 56(4), C145–C150.

Yoneda, Y. (1963). Anomalous surface reflection of X-rays. *Physical Review*, 131(5), 2010–2013.

Yulin, S.A., Feigl, T., Benoit, N., Kaiser, N. (2005). EUV/soft X-ray multilayer optics. *Proceedings of SPIE*, 5645, 289.

Zubarev, E.N., Kondratenko, V.V., Sevryukova, V.A., Yulin, S.A., Feigl, T., Kaiser, N. (2008). The structure of Mo/Si multilayers prepared in the conditions of ionic assistance. *Applied Physics A*, 90(4), 705–710.

6

Physics of Artificial Bragg Structures: Presentation of Some Phenomena

Jean-Michel ANDRÉ, Karine LE GUEN and Philippe JONNARD

LCPMR, CNRS, Sorbonne Université, Paris, France

6.1. Introduction

Other than their applications in optics, spectroscopy, etc., artificial Bragg structures used in X-ray radiation can be at the origin of various phenomena or effects, some of which have been observed in other spectral domains. Some of these phenomena or effects are described in this chapter: the Kossel effect specific to X-ray radiation, the Purcell effect initially considered in the radiofrequency domain, the recently much-studied topological effects in the context of photonic crystal optics, and finally the magnetic effects, including those related to X-ray radiation, that have burgeoned with the development of synchrotron radiation.

6.2. Kossel diffraction

According to the dynamic theory of X-ray diffraction, a characteristic emission of X-rays generated by an atom located inside a periodic medium can be diffracted by the structure itself under Bragg conditions (James 1948). This emission is subsequent to the ionization of a core orbital of an atom during irradiation by a beam of X-rays or charged particles and is due to the spontaneous recombination of the electronic cloud. The interferences due to this diffraction process result in changes in intensity of the X-ray radiation as a function of the detection angle. This very first

X-ray Radiation and Artificial Bragg Structures,
coordinated by Jean-Michel ANDRÉ and Philippe JONNARD.

interpretation of X-ray diffraction, given in 1912 by Max von Laue, was established approximately 20 years later by Walter Kossel, using the electronic excitation of a crystal, which led to the observation of so-called Kossel lines or curves (Authier 2012).

Different ionizing radiations can be used to generate Kossel lines (Langer and Daebritz 2010; Le Guen et al. 2019):

– electrons from an electron gun, a scanning electron microscope or a transmission electron microscope;

– X-ray photons generated by an X-ray tube, a plasma source or synchrotron radiation;

– fast charged particles, beams of protons or ions, from an accelerator.

The angular resolution is governed by the opening of the detector, which must be located far from the sample or collimated with a slit, in order to limit the angular opening and allow contrasted Kossel curves to be observed. In addition, the collected intensity is low, and the acquisition time can be long (a few hours). However, the configuration of the incident beam is more flexible than in classic standing wave experiments (Bedzyk and Libera 2013). This can be useful, for example, to reduce the footprint of the incident beam on the surface and thus obtain a lateral resolution on non-homogeneous samples. There is also no strict requirement for monochromaticity in the incident beam. It is sufficient for the incident energy to be higher than the binding energy of the core orbital involved in the characteristic emission observed. Kossel's experiments are also of interest when the option of irradiation by photons is not available. However, using incident X-ray radiation, the *photon-in photon-out* technique allows insulating samples (presence of a silica substrate or of dielectric layers, for example) to be analyzed without a charge effect occurring.

Kossel diffraction can be applied to all periodic materials, including crystals, but here the study is limited to periodic multilayers. Near Bragg conditions, the standing wave generated by the characteristic emission has the same period as the multilayer. Turning the sample within an angular range centered on the Bragg angle (see Figure 6.1), defined by the period of the multilayer and the wavelength of the emitted radiation, it is possible to move the nodes and antinodes of the electric field towards particular zones within the stack, for example, near an interface or the center of a layer, and thus locate the origin in terms of depth of the X-ray emission to a high degree of sensitivity (Bedzyk and Libera 2013). Therefore, Kossel diffraction is useful in determining the structural parameters (thickness, roughness and composition of the layers and any interface layers) of a periodic stack. This is a complementary tool to more widely used techniques such as X-ray fluorescence, photoelectron spectroscopy, X-ray reflectivity and transmission electron microscopy.

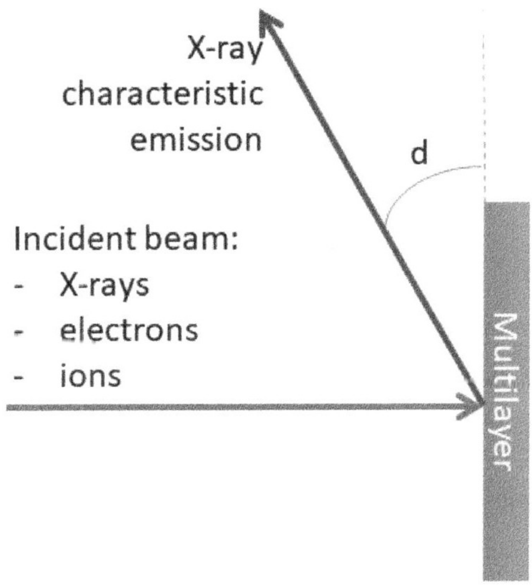

Figure 6.1. *Configuration of a Kossel diffraction experiment: the angle of detection denoted as "d" varies within the range of the Bragg angle related to the period of the multilayer and to the emitted energy; the incident beam is fixed, and the experiment is performed by turning either the sample if the detector is fixed or the detector if the sample is fixed*

Measurements can be made by turning the sample while maintaining the detector in a fixed position or with a fixed sample and a rotating detector. Detection is performed by an energy-dispersive spectrometer, generally a silicon drift detector. A spectrum is acquired for each of the detection angles within a targeted angular range. Then, the intensity measured under a characteristic peak of an element, after subtraction of the background, is traced out as a function of the detection angle. This so-called Kossel curve presents a modulation of intensity centered on the Bragg angle.

Calculating the intensity and spatial distribution of X-ray emission in Kossel diffraction requires simulation of the incident beam and of the detected X-ray radiation. We will not repeat the description of methods described in previous chapters concerning calculation of the intensity of the electric field in a multilayer structure. Calculation of the distribution with depth of the incident excitatory radiation depends considerably on the type of the radiation. For incident photons, a Beer–Lambert law is sufficient to describe attenuation of the beam, provided that we are far from the Bragg conditions. For electrons, the energy distribution is a more complicated problem. Indeed, they can undergo numerous elastic and inelastic collisions, such that a detailed description of their individual trajectories is required. The energy (and therefore, the ionization cross-section) of protons of several MeVs, such as those delivered by electrostatic accelerators, can be considered constant across the thickness of the multilayer, which is of the order of a few hundred nanometers.

We present below an example of Kossel curves obtained under photon excitation in a synchrotron center (Tu et al. 2015). The multilayers, prepared by magnetron sputtering, are the following periodic trilayers:

– Si substrate / [Mg (5.45 nm) / Co (2.45 nm) / Zr (1.50 nm)]x30 / B_4C (3.50 nm) denoted as "Mg/Co/Zr";

– Si substrate / [Mg (5.45 nm) / Zr(1.50 nm) / Co (2.45 nm)]x30 / B_4C (3.50 nm) denoted as "Mg/Zr/Co".

In other words, these multilayers only differ in the order of their deposition. The thicknesses indicated are those targeted in the preparation. The energies of the incident photons used to generate Co Lα (3d – $2p_{3/2}$ transition) and Mg Kα (2p – 1s transition) emissions are 807.6 and 1,332 eV respectively (Jonnard and Bonnelle 2011).

Figure 6.2 presents the Kossel curves measured for these two multilayers, in other words, the variation in intensity of the Co Lα and Mg Kα characteristic emissions as a function of the detection angle. On all these curves, above an increasing background, changes in intensity appear at the Bragg angles, calculated using the energy of the emissions and the multilayer period at the first and second orders of diffraction. These changes appear at smaller angles for Mg Kα emission, due to its higher energy (1,254 eV) compared to Co Lα (776 eV) emission. The curve corresponding to the Co Lα emission of Mg/Zr/Co is less contrasted than the curve for Mg/Co/Zr. Zr-on-Co interfaces (present in the stack Mg/Co/Zr) are abrupt, while Co-on-Zr interfaces (present in the Mg/Zr/Co stack) are sufficiently diffuse for it to be possible to consider both Co and Zr layers as a single mixed layer. Therefore, it is possible to adjust the Kossel curves for Mg/Co/Zr as a trilayer system, whereas for the Mg/Zr/Co system, it is necessary to model a bilayer Mg/Co_xZr_y system by introducing the compound Co_xZr_y such that $x/y = 3.5$.

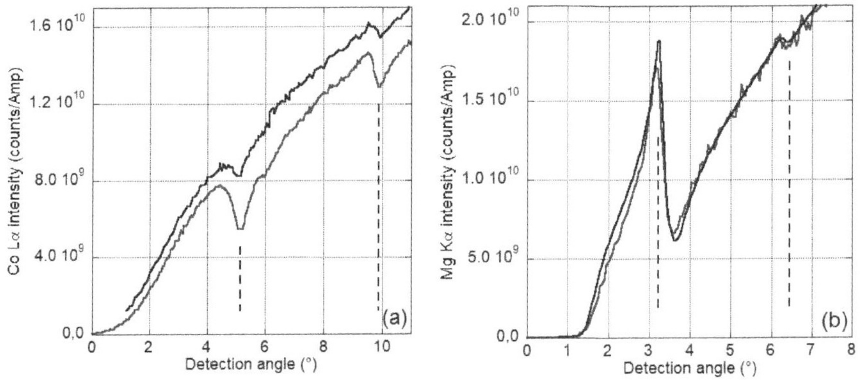

Figure 6.2. *Kossel curves of the Mg/Zr/Co (blue curve) and Mg/Co/Zr (red curve) multilayers relative to the characteristic emissions (a) Co Lα and (b) Mg Kα; the vertical bars indicate the Bragg angle at the first and second diffraction orders. For a color version of this figure, see www.iste.co.uk/andre/xrayradiation.zip*

Kossel diffraction has been applied more recently (Zhang et al. 2019) to non-periodic structures such as planar waveguides. These structures are only composed of a few layers, generally a so-called guide layer made of a light material, surrounded by two so-called cladding layers of a heavier material. The dynamics of Kossel diffraction in periodic multilayers has been studied theoretically in the case of irradiation by incident beams of ultra-short and ultra-intense X-ray radiation produced by free electron lasers (Peyrusse et al. 2020, 2021). In this case, the multilayer is both an emitter and a resonator, and the calculations show that it would be possible to observe a *stimulated X-ray emission* in the direction of the Bragg angle.

6.3. Purcell effect

In a study that has become a classic (Purcell 1946), Purcell demonstrated that it is possible to modify the spontaneous emission rate in the radio-frequency domain when the emission conditions become non-trivial. This effect, which currently bears his name, has been observed in other spectral domains, in particular the visible and infrared domains. This effect has also been observed in the X-ray radiation domain with a multilayer Bragg structure (Jonnard et al. 2002; André and Jonnard 2010). A Bragg structure composed of a periodic multilayer formed from 40 Mo/Si bilayers with period 4.8 nm and a division ratio equal to 0.33 is bombarded with monoenergetic electrons with energy of 6 keV. The intensity of the Si Kα emission at 0.713 nm is recorded as a function of the angle formed by the surface of the multilayer and the direction of the incident electron beam, known as the detection angle. The line is isolated using a wavelength-dispersive spectrometer as a monochromator. Figure 6.3 presents the experimental variation in the emission intensity of the Si Kα line, along with the calculated density of mode $\rho(\omega)$ (DOM), and with a simulation of the power emitted taking into account instrumental broadening (André and Jonnard 2011).

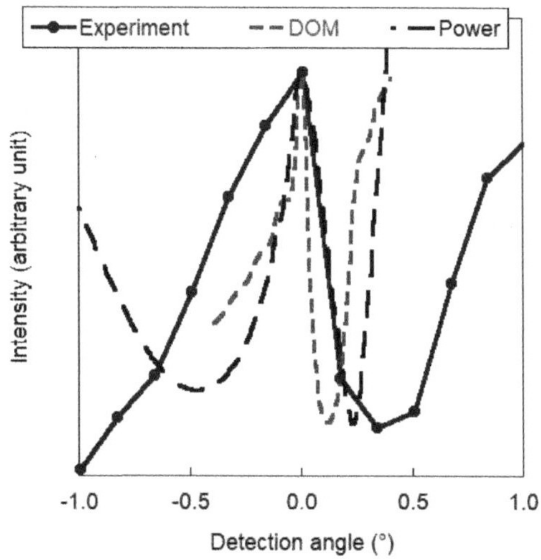

Figure 6.3. *Intensity of the Si Kα emission of an Mo/Si multilayer as a function of the angle of detection compared to the density of mode and to a simulation of the power; the angles are given with respect to the Bragg angle at first diffraction order. For a color version of this figure, see www.iste.co.uk/andre/xrayradiation.zip*

These curves show a significant variation, in particular a peak located at angle 0°, corresponding to the Bragg angle, followed by a fall in intensity which is not observed with a bulk silicon target. These structures, which are associated with the periodic structure at Bragg resonance, have been interpreted as the signature of the Purcell effect and modeled in the linear regime using variations on the spontaneous emission rate τ given by the Fermi golden rule (Cohen-Tannoudji et al. 1997):

$$\tau = \frac{2\pi}{\hbar}\rho(\omega)\,|\langle f|H|i\rangle|^2\,\delta(E_i - E_f) \qquad [6.1]$$

where i and f respectively denote the initial energy state E_i and the final energy state E_f of the transition corresponding to the spontaneous emission. H is the overall Hamiltonian of the system, comprising the emitting atom, the radiation and the

Bragg structure which constitutes the emitting cavity. This rate can be calculated in the model of the harmonically oscillating dipole at the resonance frequency of the transition $\omega_0 = \frac{|E_f - E_i|}{\hbar}$. The average power emitted in the stationary regime P_ω is expressed using the optical DOMs (see Chapter 2), $\rho(\omega)$ and the eigenmodes of the field $\boldsymbol{a_k}$:

$$P_\omega = C\,\rho(\omega)\,|\boldsymbol{a_k}|^2 \quad [6.2]$$

where C is a constant, the value of which is not relevant in the case of a qualitative description. The DOM can be calculated using the method of Bendickson et al. (1996) presented in Chapter 2. The eigenfields only depend on the variable z according to the axis of stratification: $\boldsymbol{a_k} \equiv a_k(z)$ and obey the Helmholtz equation which is written with the usual notations.

$$\ddot{a}_k(z) + \frac{\omega_k}{c}\,\varepsilon(z) a_k(z) = 0 \quad [6.3]$$

of which the solution in the pth layer of type j is written (omitting the index k) as:

$$a_j^{(p)}(z) = exp(i\,p\,K\,d)\left[A_j\,exp\left(i\,k_{\perp j}(z-d)\right) + B_j\,exp\left(-i\,k_{\perp j}(z-d)\right)\right] \quad [6.4]$$

where $p\,d < z < (p+1)d$. K is the Bloch number given by the dispersion equation (see Chapter 2), and $k_{\perp j}$ is the normal component of the wave vector in the medium j. The coefficients A_j and B_j are obtained from the continuity conditions at the limits.

Figure 6.3 shows the DOM and the power P_ω calculated by the above theory in comparison with the experiment, as a function of the angle of detection. The DOM presents a peak that coincides with the experimental peak and the peak emitted power. This peak is located at the edge of the stopband of the Bragg structure. This exaltation of the DOM explains the increase in intensity of the spontaneous emission of the Si Kα line inside the stopband. The DOM descends below 1 (value corresponding to the DOM in a vacuum), which expresses the reduction in intensity of the emission. This behavior is typical of the Purcell effect in a one-dimensional photonic crystal and is closely linked to the Kossel diffraction presented in section 6.2 (André et al. 2015).

6.4. Bragg–Raman effect

Kossel diffraction is based on the spontaneous emission of an ionized atom. The Bragg–Raman effect observed by André et al. (2005) in the field of X-UV radiation is a rarer phenomenon in the sense that its occurrence requires more restrictive conditions. This effect originates in inelastic diffusion (Raman effect) of incident radiation by an atomic species constituting the crystal, with the scattered radiation diffracted under Bragg conditions by the crystalline lattice. In the phenomenon described in the reference cited above, the Bragg structure is a periodic multilayer composed of La/B_4C bilayers. The quasi-monochromatic X-UV radiation, obtained from synchrotron radiation, is scattered inelastically by the *5p* electrons of lanthanum. The experimental scheme is shown in Figure 6.4(a).

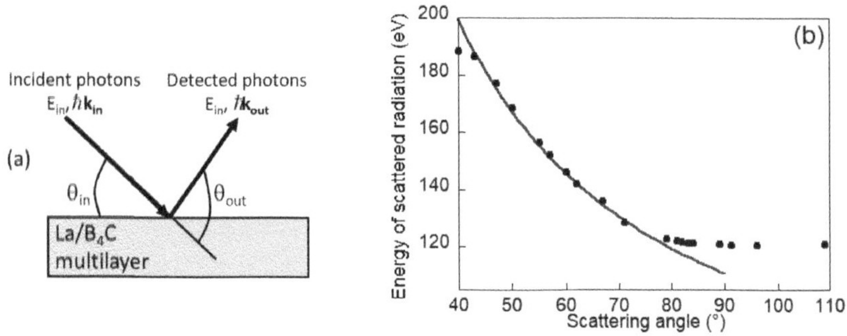

Figure 6.4. *(a) Scheme of the Bragg–Raman experiment; (b) energy E_{out} of the radiation scattered during the Bragg–Raman effect, as a function of the scattering angle θ_{out}; dots: experiment; solid line: kinematic model*

The phenomenon can be modeled using a kinematic approach by writing the energy and momentum conservation laws. We write:

– for energy conservation:

$$E_{in} = E_{out} + E_X \qquad [6.5]$$

with E_{in} being the energy of the incident photon, E_{out} being the energy of the scattered photon and E_X being the energy associated with the Raman scattering process;

–for momentum conservation:

$$\hbar\, \boldsymbol{k_{in}} = \hbar\, \boldsymbol{k_{out}} + \hbar\, \boldsymbol{g} + \boldsymbol{q_X} \qquad [6.6]$$

where $\hbar\, \boldsymbol{k}_{in}$ and $\hbar\, \boldsymbol{k}_{out}$ are the momenta associated with the incident and scattered photon, respectively, $\boldsymbol{q}_X$ is the momentum associated with the Raman scattering process and $\hbar\, \boldsymbol{g}$ is the momentum related to the crystalline lattice ($|\,\boldsymbol{g}| = g$). By associating these two conservation equations, we can obtain the resonant excitation energy $\widetilde{E_{in}}$ given by the relationship:

$$\widetilde{E_{in}} = \frac{E_X}{2} + \frac{\sqrt{2\,c^2\,q^2\,(1-cos(\theta_{out})) - {E_X}^2\,sin^2(\theta_{out})}}{2\,(1-cos(\theta_{out}))} \quad [6.7]$$

with:

$$q^2 = (\hbar\,\boldsymbol{g} + \boldsymbol{q}_X)^2 \quad [6.8]$$

where θ_{out} is the scattering angle. The previous relationship is the generalization of the standard Bragg law to the case of inelastic photon scattering (Raman). It is easy to verify that it returns to the standard law for the elastic case $E_X = 0$. From this relationship, the experimental curve $\widetilde{E_{in}}$ can be adjusted as a function of θ_{out} as shown in Figure 6.4(b), provided that the approximate value of 20 eV (corresponding to the $5p^5$ configuration of lanthanum in the physical and chemical state within the multilayer structure) is taken for E_X, and the value given by $\frac{\pi}{2} + \theta_{out} - \theta_{in}$ is used for the angle θ_X between $\boldsymbol{q}_X$ and $\boldsymbol{g}$.

The phenomenon has not yet revealed all its secrets, insofar as to interpret it, it is necessary to allow the existence of a momentum $\boldsymbol{q}_X$ of physical characteristics that are not yet fully understood (and that we have stated to be photonic: $\frac{E_X}{c} = q_X$) and to impose the constraint that the angle θ_X should correspond to $\frac{\pi}{2} + \theta_{out} - \theta_{in}$.

6.5. Optical and topological states – X-ray quantum optics

In practice, standard artificial Bragg structures present themselves as periodic or quasi-periodic stacks with limited dimensions. They can be considered as truncated one-dimensional photonic crystals. Following the truncation, the surface states that are electromagnetic excitations are likely to appear between another semi-infinite periodic medium or another semi-infinite bulk medium. They are located at the interface, in contrast to the Bloch waves, which extend throughout the periodic medium. They were first studied in the 1930s in the context of the physics of solids in the works of Tamm (1932) and Shockley (1939) and were then studied again from the 1960s onwards in the context of optics of periodically stratified media under different denominations: Bloch electromagnetic surface states, Tamm optics, etc.

Since the 1980s, topology has entered the domain of physics of condensed matter, then the domain of photonic crystals via the analogy between these two domains, and topological edge states were discovered at the interface between a semi-infinite periodic optical medium and another semi-infinite homogeneous or periodic optical medium. The concept of a topological edge state is not trivial. An introduction to this concept can be found in the case of one-dimensional photonic crystals in Chapter 9 of a work by André (2023).

A large number of studies have been carried out on edge states in the long wavelength domain, in particular the optical domain (Yeh et al. 1978), but they have only very recently been considered in the X-ray radiation domain by Huang et al. (2019). The heterostructure, designed by these authors to perform and observe the topological edge states in the X-ray radiation domain, consists of a structure denoted as PC_A comprising 10 unit cells (WCW)[1], in which each layer has a thickness of 1.5 nm, joined to a symmetrical structure denoted as PC_B (CWC), with mirror symmetry with respect to the junction plan as shown in Figure 6.5(a). A signature of the topological edge state is the presence of a dip in the reflectance curve of the heterostructure $PC_A + PC_B$ which is observed in the hard X-ray radiation domain at 8.04 keV, as shown in Figure 6.5(b).

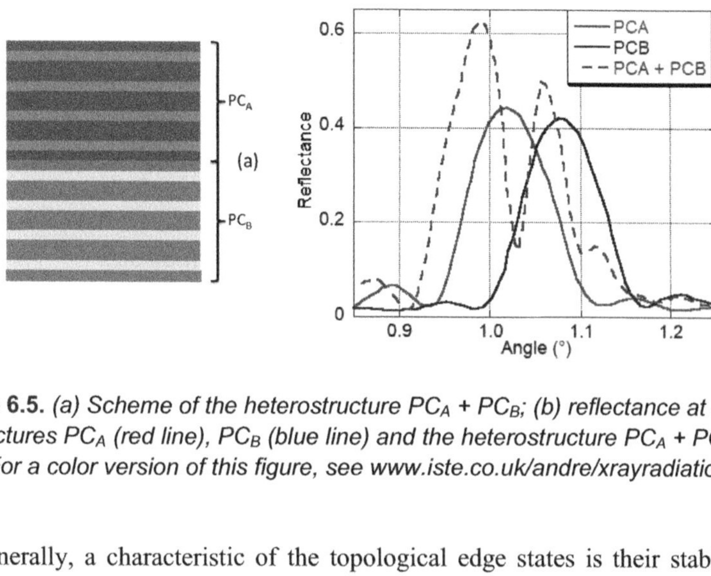

Figure 6.5. *(a) Scheme of the heterostructure $PC_A + PC_B$; (b) reflectance at 8.04 keV of structures PC_A (red line), PC_B (blue line) and the heterostructure $PC_A + PC_B$ (green line). For a color version of this figure, see www.iste.co.uk/andre/xrayradiation.zip*

Generally, a characteristic of the topological edge states is their stability with regard to disorder. This property is verified in this work, in this case the stability with respect to a disturbance of the thickness of the layers: the dip observed in the reflectance of the heterostructure is maintained despite a low level of disorder in the thickness of the layers. Another important result recorded in this work is the observation of Rabi splitting: a fine layer of ^{57}Fe is added at the junction between PC_A and PC_B; the element ^{57}Fe is a Mössbauer isotope, which presents a system on two levels with a transition frequency equal to 3.48 10^{18} Hz, that is, 14.4 keV. Rabi

1. The symbols W and C denote respectively the elements tungsten and carbon.

splitting occurs subsequent to the presence of the topological edge state, which leads to local exaltation (at the junction where the ^{57}Fe layer is located) of the electromagnetic field which oscillates at the Rabi transition frequency. This splitting manifests itself in the appearance of two dips in the reflectance curve.

The isotope ^{57}Fe has already been used extensively in the field of X-ray nuclear optics, which is in a phase of considerable development, in particular from an experimental point of view with the development of sources of hard X-ray radiation, namely fourth-generation synchrotrons and free electron lasers. Indeed, the energy domain in which this field of physics has been developed is in the high energy regime (hard X-rays), more particularly at 14.412 keV, which corresponds to a nuclear magnetic resonance energy of iron isotope 57. This is a very narrow resonance (Mössbauer transition with a natural width of 4.66 neV!) compared to the resonances seen in the domain of electronic transitions.

Therefore, many studies have been carried out over the last 30 years in a domain of optics with high energy radiation, sometimes known as "X-ray quantum optics", involving "nuclear optical lattices" composed of ^{55}Fe/^{57}Fe multilayers; in particular the works of Chumakov et al. (1993) and Röhlsberger et al. (1993). The authors of the latter work report that the high exaltation of the radiative scattering channel leads to a fast time response at the Bragg peak of ^{57}Fe resonance (decay time 4 ns!) and highlight that this nuclear Bragg structure could constitute a very narrow bandwidth monochromator for the synchrotron radiation.

It is important to point out that the observation of phenomena corresponding to nuclear magnetic scattering is very difficult because their intensity is several orders of magnitude lower than for electronic scattering and requires specific instrumentation using filters combining polarizer and analyzer. Most of the experimental works performed on this system relate to measurement of Bragg diffraction by reflectometry. A study published in 2016 (Haber et al. 2016) tackled the spectral aspect and revealed the strong coupling, via Rabi doubling, between the X-ray radiation interacting with the core inside the cavity. Together, these works pave the way to new studies of quantum phenomena in cavities in the X-ray radiation range.

6.6. Magnetism

The physics related to the association between magnetism, X-ray radiation and artificial Bragg structures primarily involve magnetic-optical effects, which are the Kerr effects, the Faraday effect and dichroism, discovered at the end of the 19th century. The Kerr and Faraday effects manifest themselves respectively as a polarization rotation during reflection or transmission of the radiation. These effects are often more marked in the domain of X-ray radiation than in the domain of the longest wavelengths (Froman 1932; Kuneš et al. 2001). The Kerr effect in the domain of X-ray radiation, sometimes also referred to by the term X-MOKE, is in particular extensively used to characterize the magnetic properties of a material (Lee et al. 2003; Kim et al. 2005).

The interaction of polarized X-ray radiation with a magnetic material can give rise to a phenomenon known as magnetic dichroism that consists of a differentiated response of the material to the type of radiation polarization. We distinguish between X-ray magnetic circular dichroism (XMCD) and X-ray magnetic linear dichroism (XMLD). XMCD concerns circular polarized radiation and is the dominant effect with ferri- or ferro-magnetic materials. It can be observed either by:

– reversing the magnetization of the sample under fixed polarization of the radiation;

– modifying the circular polarization (change in helicity) while conserving the direction of magnetization;

– reversing the spin–orbit coupling under constant polarization and magnetization.

For these observations, the change in circular polarization helicity is generally obtained by suitably adjusting the parameters of an undulator on a synchrotron. This is preferable to reversal of the magnetization, which can give rise to remanence effects. The second phenomenon, XMLD, concerns linearly polarized radiation primarily in interaction with antiferromagnetic (AFM) materials.

These different phenomena can be observed either in transmission mode via measurement of the transmittance or in reflection mode via measurement of the reflectance. The dichroism in X-ray fluorescence mode has also been highlighted (Hague et al. 1993). The magnetic dichroism of the X-ray radiation is more marked near certain absorption thresholds of the elements. We then refer to resonant magnetic scattering. This resonant exaltation is particularly marked by spin–orbit splitting at the L and M thresholds of $3d$ (transition metals), $4f$ (rare-earth elements) and $5f$ (actinides) elements.

The dichroism is generally quantified by a quantity D known as the asymmetry ratio (dichroic) and it is defined by:

$$D = \frac{I^+ - I^-}{I^+ + I^-} \qquad [6.9]$$

where $I^{\pm}$ is the intensity transmitted, reflected or scattered (emission of fluorescence) for a positive (+) or negative (–) helicity of the radiation. Figure 6.6 shows a resonant XMCD effect at the $L_{2,3}$ thresholds of iron (Fischer and Ohldag 2015).

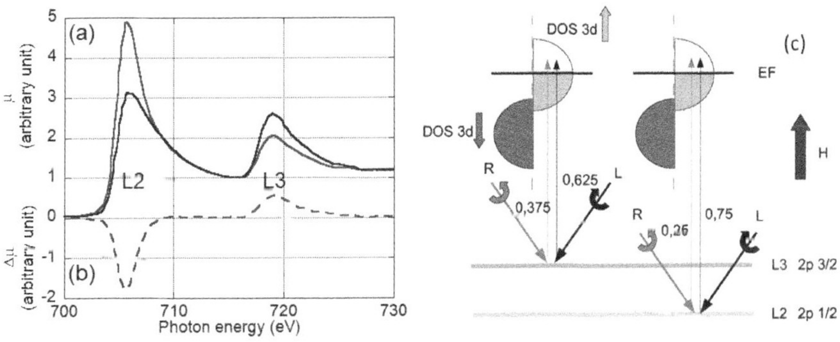

Figure 6.6. *Resonant XMCD at the $L_{2,3}$ thresholds of iron. For a color version of this figure, see www.iste.co.uk/andre/xrayradiation.zip*

COMMENT ON FIGURE 6.6.– *(a) Resonant absorption spectra of iron at the $L_{2,3}$ thresholds for two relative orientations of the magnetization of the sample and polarization of the radiation (parallel orientation red line; antiparallel orientation blue line); (b) XMCD signal deduced from the spectra presented in (a); (c) principle of XMCD for a 3d transition metal: the photoelectron transits with a different probability (0.75, ...) towards the unoccupied states above the Fermi level (E_F) depending on the starting level L_2 or L_3, the spin up (blue arrow pointing upwards) or spin down (red arrow pointing down) type and the helicity (left L, right R) of the circular polarized radiation.*

Whatever the mode of observation, the basic mechanism of dichroism is the process of photo-absorption with a cross-section that is proportional to the probability of transition $\Gamma_{i\to f}$ from the initial state i to the final state f given by the Fermi golden rule:

$$\Gamma_{i\to f} = \frac{2\pi}{\hbar}|\langle f|H_{int}|i\rangle|^2\,\rho_{in}(E) \qquad [6.10]$$

where H_{int} is the Hamiltonian for interaction between the X-ray radiation and the atom and $\rho_{in}(E)$ is the density of unoccupied states as a function of the energy E. In the case of fluorescence, which can be seen as a two-stage process (absorption and then emission), the density of the occupied states $\rho_{oc}(E)$ is also involved via the emission process. In this respect, the dichroism in X-ray fluorescence mode constitutes a probe of the states occupied with magnetic selectivity (Strange et al. 1993).

The Fermi rule allows XMCD spectra to be interpreted, such as those presented in Figure 6.6. The spin–orbit coupling separates the electronic level $2p$ into two sub-levels: one corresponding to the absorption threshold L_2 ($2p_{1/2}$) with the quantum number j equal to 1/2 and antiparallel orbital and spin moments, and the other corresponding to the absorption threshold L_3 with the number j equal to 3/2 ($2p_{3/2}$) and parallel orbital and spin moments. The photoelectron is polarized in terms of spin. It has a spin *down* or spin *up*[2] depending on the antiparallel or parallel configuration of the moments (orbit and spin) by virtue of conservation of the angular moment but with a different probability depending on the level L_2 or L_3 and the helicity of the radiation. In the model with one electron (Stöhr and Wu 1994; Stöhr 1995), it occurs that, for a circular polarized radiation with left helicity, a spin *down* photoelectron is emitted with probability 3/4 from level L_2 and a spin *up* photoelectron is emitted with probability 5/8 from level L_3 (see Figure 6.6(c)). Consequently, the photoelectron is likely to undergo a transition towards unoccupied states above the Fermi level but with a different probability depending on the *down* or *up* direction of the spin states (see Figure 6.6(c)), which explains the XMCD. The reasoning is generalized for the other thresholds.

To model the interaction of the X-ray radiation with magnetic materials, we introduce an additional term f_{res} into the expression of the atomic scattering factor presented in Chapter 2. Hannon et al. (1988) have written the term f_{res} in the dipolar approximation in the form of a development in the power of $\boldsymbol{u}$, the unit vector of quantification:

$$f_{res} \propto (\boldsymbol{e}_{sc}^{*}.\boldsymbol{e}_{in})\,F_0 \;-i\,\{(\boldsymbol{e}_{sc}^{*}\wedge\boldsymbol{e}_{in}).\boldsymbol{u}\,F_1 + \;(\boldsymbol{e}_{sc}^{*}.\boldsymbol{u})(\boldsymbol{e}_{in}.\boldsymbol{u})(2\,F_0\} + \cdots \qquad [6.11]$$

with $\boldsymbol{e}_{in}$ and $\boldsymbol{e}_{sc}$ being the unit vectors associated with the electric field of the incident (*in*) and scattered (*sc*) radiation. The terms F_0 and F_1 are written using quantities denoted by F_{11} and $F_{1\bar{1}}$ as follows:

$$F_0 = F_{11} + F_{1\bar{1}};\; F_1 = F_{11} - F_{1\bar{1}} \qquad [6.12]$$

2. A photoelectron produced by absorption of linear or circular polarized radiation has three polarization components generally located with respect to the scattering plan. These can be obtained using the theory of quantum mechanics (Cherepkov 1983).

where F_{1m} are quantities proportional to the absorption probability of a photon with a change m in the projection of its angular moment on the quantification axis. These quantities vary anomalously, in the sense of anomalous scattering, in resonance regions, in other words, at the absorption thresholds, as shown in Figure 6.7 regarding iron at the $L_{2,3}$ thresholds (Elzo et al. 2012).

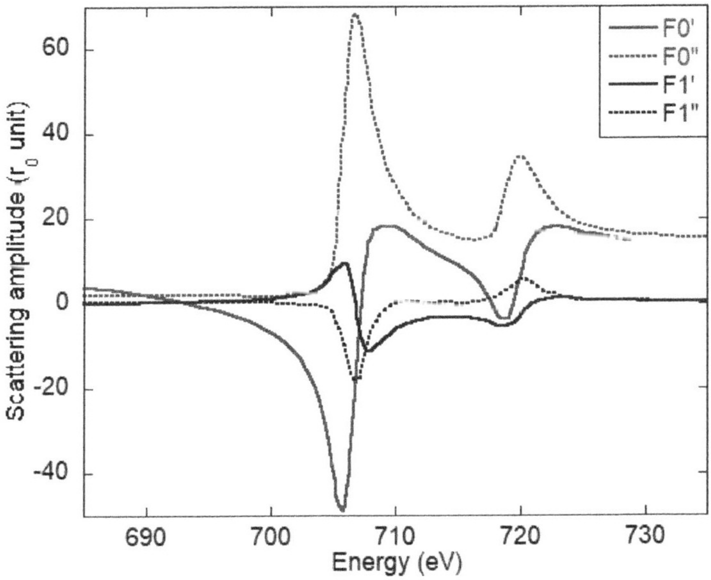

Figure 6.7. *Variation, as a function of the radiation energy, of the real part denoted ' (continuous lines) and the imaginary part denoted " (dotted lines) of the terms F_0 and F_1 for iron near the $L_{2,3}$ thresholds. For a color version of this figure, see www.iste.co.uk/andre/xrayradiation.zip*

Using the scattering factor with its additional magnetic term f_{res}, an optical[3] constant can be constructed, taking into account the magnetic effects, which are introduced into the optical calculation models. Therefore, various models generalizing the matrix approach presented in Chapter 2 have been developed for the calculation of reflectance, transmittance, etc., for multilayer structures using optical constants with magnetic corrections (Zak et al. 1991; Ishimatsu et al. 1999; Elzo et al. 2012; Verna et al. 2021).

3. The optical constant will generally be a tensor and will be expressed using a square matrix with dimension 3: $\bar{\bar{\epsilon}} = \begin{pmatrix} \epsilon_{xx} & \epsilon_{xy} & \epsilon_{xz} \\ \epsilon_{yx} & \epsilon_{yy} & \epsilon_{yz} \\ \epsilon_{zx} & \epsilon_{zy} & \epsilon_{zz} \end{pmatrix}$.

From an experimental point of view, resonant magnetic scattering under its different modes (transmission, for example) has become a powerful tool for studying nano-magnetism and determining magnetization profiles. The first studies mainly involved bulk materials, rare-earth elements and actinides at their $L_{2,3}$ and $M_{4,5}$ thresholds (McWhan 1994); then, owing to the admissible beams in polarization in the domain of soft X-ray radiation, many investigations were carried out on the $L_{2,3}$ thresholds of transition metals (Kao et al. 1994; Tonnerre et al. 1995). The reflection mode is prioritized for thick-system, periodic or non-periodic multilayers.

It is important to highlight that the presence of an interface can modify the magnetic characteristics of the materials located on either side of the material: for example, the appearance of magnetization induced by the neighboring material or of a magnetic dead zone. This problem constitutes a key subject of study for applications of magnetic materials, and in this context, using the reflection mode, valuable information is added about buried interfaces (Verna et al. 2021). The periodic case is particularly interesting because the reflection in Bragg conditions offers fruitful possibilities to be exploited. We present some of them below.

The first possibility is based on the simplicity of analysis of Bragg diffraction with a low value for the transfer moment, in relation to a specular measure in a non-interferential situation (order 0). Indeed, in this case of Bragg diffraction, a kinematic approach not taking into account the multiple scattering is generally sufficient for evaluation of the diffracted intensity: from the module of the structure factor involving the magnetic scattering factor and the measured diffracted intensity, it is possible to return to this scattering factor and therefore to the magnetic characteristics. Figure 6.8 presents the relation between dichroic asymmetry and the diffracted intensity in longitudinal mode (XMCD) in the first Bragg order in the absorption region $L_{2,3}$ for the nickel in a periodic multilayer $Fe_{19}Ni_{81}/Ag$.

A second possibility is based on the standing waves system which establishes itself in the Bragg regime (see Figure 2.3): by playing on the geometry, for example, by modifying the angle of incidence around the Bragg angle, the "centers" of this standing wave can be moved, and therefore, the intensity of the electric field can be modified locally. A spatial probe is thus added to the chemical probe constituting resonant scattering. This combined technique, XMCD + standing waves, referred to in the literature under the acronym MXSW (magnetic X-ray standing waves) (Zegenhagen and Kazimirov 2013), was applied to highlight the local ferromagnetism of the carbon layers induced by the iron atoms adjacent to the interfaces in an Fe/C periodic multilayer (Mertins et al. 2004).

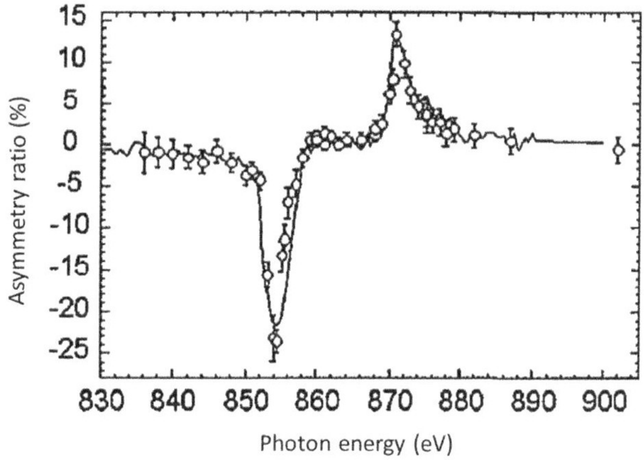

Figure 6.8. *Relation between the dichroic asymmetry and the diffracted intensity in longitudinal mode (XMCD) at the first Bragg order in the absorption region $L_{2,3}$ for the nickel in a periodic multilayer: points, experiment; solid line, kinetic calculation (Raoux et al. 2000)*[4]

Another example of the application of this technique is the demonstration of a dichroism phenomenon of *X-ray fluorescence* in an Mg/Co periodic multilayer (Jonnard et al. 2018). Another possibility is based on the sensitivity of the position of the Bragg peak, in terms of angle, frequency or wavelength, at the value of the optical index (see equation [2.25]). Therefore, the magnetic effects modify the optical constant value, which only takes into account the electronic contribution as seen above. A magnetic shift of the Bragg peak ensues with respect to the uncorrected position. By measuring this shift, we can return to the tensorial dielectric constant $\bar{\bar{\epsilon}}$. We obtain, under certain symmetry conditions:

$$\epsilon_{yz} = -\epsilon_{zy} = (\epsilon^{+} - \epsilon^{-})/2$$

$$\epsilon_{xx} = \epsilon_{yy} = \epsilon_{zz} = (\epsilon^{+} + \epsilon^{-})/2$$

$$\epsilon_{xz} = \epsilon_{zx} = \epsilon_{xy} = \epsilon_{yx} = 0$$

$$\epsilon^{\pm} = (1 - \delta^{\pm} - i\,\beta^{\pm})^2 \qquad [6.13]$$

where $\delta^{\pm}$ and $\beta^{\pm}$ are respectively the decrement from a unit in the real part of the refraction index and $\beta^{\pm}$ is the imaginary part of this index; the term $\pm$ indicates the two opposing orientations according to the magnetization axis $\boldsymbol{x}$. This method was

4. See doi: 10.12693/APhysPolA.98.483.

used by Sacchi et al. (1998) for a multilayer Bragg structure made up of 40 Fe/V bilayers. Figure 6.9 presents the results obtained using this approach for iron indices.

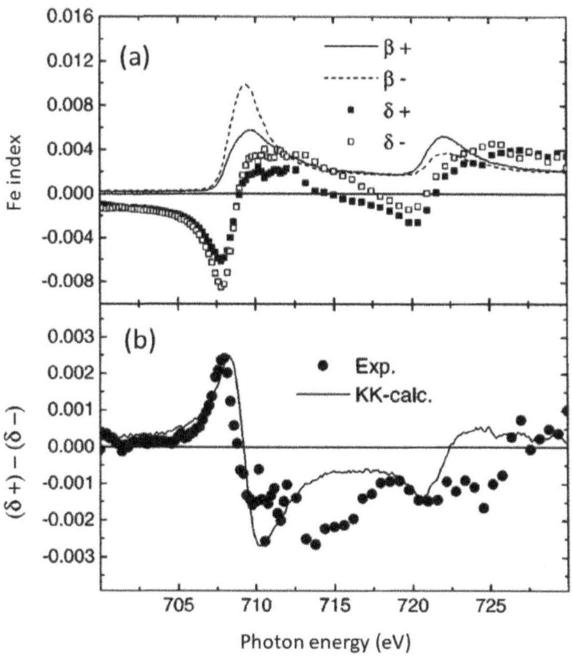

Figure 6.9. *(a) Spectrum of the decrement from a unit in the optical index δ and its imaginary part β for iron in the absorption region $L_{2,3}$ according to the orientation +/- of magnetization; (b) dichroism on δ: $\delta^+ - \delta^-$ (dots: experiment) compared to the calculated values (continuous line) with the Kramers–Kronig transformation of the imaginary part $\beta^+ - \beta^-$ (Sacchi et al. 1998)*[5]

It has been demonstrated that an AFM Bragg structure can present double periodicity: the first so-called structural with a period d corresponding to the geometric periodicity resulting from fabrication and the second corresponding to a super-lattice with period $D = 2d$ resulting from the magnetic effect AFM. Resonant magnetic X-ray scattering (RMXS) experiments have shown that diffraction curves (reciprocal scan) reveal magnetic AFM Bragg peaks appearing at half orders ($L = n/2$ ($n \in \mathbb{N}$), where L is the reciprocal lattice unit) next to structural Bragg peaks appearing at integer orders ($L = n$) (Nefedov et al. 2005a, 2005b; Schick et al. 2021).

5. See doi: 10.1103/PhysRevLett.81.1521.

In Schick et al. (2021), the structure is made up of Fe/Cr bilayers with thickness $d = 1.8$ nm and the super-period D contains two ferromagnetic layers of iron. Figure 6.10 shows a *reciprocal scan*, which presents a structural Bragg peak of order 1 ($L = 1$) and a magnetic AFM Bragg peak at $L = 1/2$ (half order) *when the energy is resonant (level L3 of iron).*

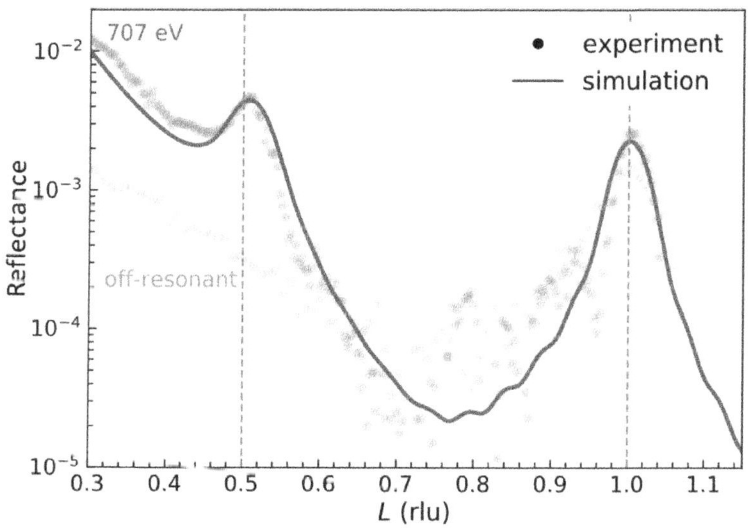

Figure 6.10. *Diffraction pattern (reciprocal scan), expressed in unit of reciprocal lattice L, of a periodic structure formed from Fe/Cr bilayers. For a color version of this figure, see www.iste.co.uk/andre/xrayradiation.zip*

COMMENT ON FIGURE 6.10.– *The peak at L = 1 is the structural Bragg peak whereas the peak at L = 1/2 is the magnetic AFM Bragg peak which appears at the resonant energy (707 eV corresponding to the L_3 threshold for iron); the orange dots present the case excluding resonance at 680 eV. The solid blue line on the graph is a simulation taking into account the magnetic effects; reproduced with permission from Schick et al. (2021)[6]. © The Optical Society.*

A magnetic AFM super-lattice was also observed in a monocrystalline homogeneous film (not artificially periodic) of EuTe (Holldack et al. 2010). The AFM structure consists of planes (111) ordered according to ferromagnetization, alternating with the magnetization directions [111], such that the magnetic Bragg peaks appear at half orders, in other words for $L = n/2$ ($n \in \mathbb{N}$). In the last two works

6. See doi: 10.1364/OPTICA.435522.

cited above, temporal studies of magnetization were performed. Indeed, based on the works of Beaurepaire et al. (1996), who revealed an ultra-rapid demagnetization effect (at the scale of femtoseconds), the theme of dynamic analysis of magnetization has been developed considerably. These works are possible thanks to the availability of sources of X-ray radiation that are pulsed and turnable in terms of energy (synchrotron radiation in slicing mode, XFEL and also HHG sources), which perform the probe function, whereas magnetization is induced by a laser pulse in the infrared domain, which plays the role of a pump, as shown in Figure 6.11(a). Figure 6.11(b) shows the dynamic curve of the change in magnetization for the EuTe film considered below.

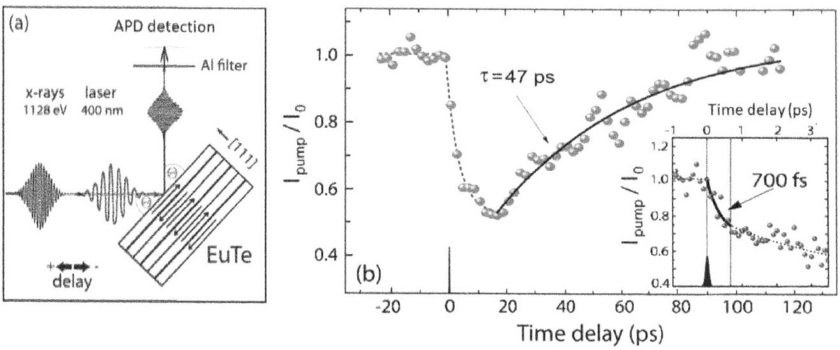

Figure 6.11. *Changes in magnetization over time with an EuTe film probed by a pulse of X-rays*

COMMENT ON FIGURE 6.11.– *(a) Scheme of the measurement device with the changes over time in magnetization of an AFM film of EuTe: an infrared laser pulse (400 nm) induces magnetization followed after a time delay by a pulse of X-rays which probes the process; (b) change over time in the magnetization, as a function of the time delay (in the smaller frame, the first part of the curve) according to Holldack et al. (2010)*[7].

6.7. References

André, J.-M. (2023). *Cristaux et quasi-cristaux photoniques unidimensionnels : principe et théorie*. Auto-édition JMA, Amboise.

7. See doi: 10.1063/1.3474612.

André, J.-M. and Jonnard, P. (2010). X-ray spontaneous emission control by 1-dimensional photonic bandgap structure. *The European Physical Journal D*, 57(3), 411–418.

André, J.-M. and Jonnard, P. (2011). Contrôle de l'émission spontanée de rayonnement X au moyen d'un cristal photonique mono-dimensionnel. In *UVX 2010 – 10ème Colloque sur les Sources Cohérentes et Incohérentes UV, VUV et X; Applications et Développements Récents*. EDP Sciences, Paris, 163–167 [Online]. Available at: https://uvx.edpsciences.org/articles/uvx/abs/2011/01/uvx2011_022/uvx2011_022.html [Accessed 13 March 2023].

André, J.-M., Jonnard, P., Bonnelle, C., Filatova, E.O., Michaelsen, C., Wiesmann, J. (2005). X-ray Raman scattering with Bragg diffraction in a La-based superlattice. *Optics Communications*, 255(4/6), 267–271.

André, J.-M., Jonnard, P., Guen, K.L., Bridou, F. (2015) Kossel diffraction and photonic modes in one-dimensional photonic crystal. *Physica Scripta*, 90(8), 085503.

Authier, A. (2012). Optical properties of X-rays – dynamical diffraction. *Zeitschrift für Kristallographie*, 227(1), 36–51.

Beaurepaire, E., Merle, J.C., Daunois, A., Bigot, J.Y. (1996). Ultrafast spin dynamics in ferromagnetic nickel. *Physical Review Letters*, 76(22), 4250–4253.

Bedzyk, M.J. and Libera, J.A. (2013). X-ray standing wave in multilayers. In *The X-Ray Standing Wave Technique: Principles and Applications*, Zegenhagen, J. and Kazimirov, A. (eds). World Scientific, Singapore, 122–131. doi: 10.1142/9789812779014_0007 [Accessed 30 October 2013].

Bendickson, J.M., Dowling, J.P., Scalora, M. (1996). Analytic expressions for the electromagnetic mode density in finite, one-dimensional, photonic band-gap structures. *Physical Review E*, 53(4), 4107–4121.

Cherepkov, N.A. (1983). Spin polarization of atomic and molecular photoelectrons. In *Advances in Atomic and Molecular Physics*, Bates, D. and Bederson, B. (eds). Academic Press, Cambridge, 395–447 [Online]. Available at: https://www.sciencedirect.com/science/article/pii/S0065219908602584 [Accessed 11 April 2023].

Chumakov, A.I., Smirnov, G.V., Baron, A.Q.R., Arthur, J., Brown, D.E., Ruby, S.L., Brown, G.S., Salashchenko, N.N. (1993). Resonant diffraction of synchrotron radiation by a nuclear multilayer. *Physical Review Letters*, 71(15), 2489–2492.

Cohen-Tannoudji, C., Diu, B., Laloë, F. (1997). *Mécanique quantique II*. Editions Hermann, Paris.

Elzo, M., Jal, E., Bunau, O., Grenier, S., Joly, Y., Ramos, A.Y., Tolentino, H.C.N., Tonnerre, J.M., Jaouen, N. (2012). X-ray resonant magnetic reflectivity of stratified magnetic structures: Eigenwave formalism and application to a W/Fe/W trilayer. *Journal of Magnetism and Magnetic Materials*, 324(2), 105–112.

Fischer, P. and Ohldag, H. (2015). X-rays and magnetism. *Reports on Progress in Physics*, 78(9), 094501.

Froman, D.K. (1932). The Faraday effect with X-rays. *Physical Review*, 41(6), 693–700.

Haber, J., Schulze, K.S., Schlage, K., Loetzsch, R., Bocklage, L., Gurieva, T., Bernhardt, H., Wille, H.-C., Rüffer, R., Uschmann, I. et al. (2016). Collective strong coupling of X-rays and nuclei in a nuclear optical lattice. *Nature Photonics*, 10(7), 445–449.

Hague, C.F., Mariot, J.-M., Strange, P., Durham, P.J., Gyorffy, B.L. (1993). Observation of magnetic circular dichroism in Fe L2,3 X-ray-fluorescence spectra. *Physical Review B*, 48(5), 3560–3562.

Hannon, J.P., Trammell, G.T., Blume, M., Gibbs, D. (1988). X-ray resonance exchange scattering. *Physical Review Letters*, 61(10), 1245–1248.

Holldack, K., Pontius, N., Schierle, E., Kachel, T., Soltwisch, V., Mitzner, R., Quast, T., Springholz, G., Weschke, E. (2010). Ultrafast dynamics of antiferromagnetic order studied by femtosecond resonant soft X-ray diffraction. *Applied Physics Letters*, 97(6), 062502.

Huang, Q., Guo, Z., Feng, J., Yu, C., Jiang, H., Zhang, Z., Wang, Z., Chen, H. (2019). Observation of a topological edge state in the X-ray band. *Laser & Photonics Reviews*, 13(6), 1800339.

Ishimatsu, N., Hashizume, H., Hamada, S., Hosoito, N., Nelson, C.S., Venkataraman, C.T., Srajer, G., Lang, J.C. (1999). Magnetic structure of Fe/Gd multilayers determined by resonant X-ray magnetic scattering. *Physical Review B*, 60(13), 9596–9606.

James, R.W. (1948). *The Optical Principles of the Diffraction of X-rays*. G. Bell and Sons, London.

Jonnard, P. and Bonnelle, C. (2011). Cauchois and Sénémaud tables of wavelengths of X-ray emission lines and absorption edges. *X-Ray Spectrometry*, 40(1), 12–16.

Jonnard, P., André, J.-M., Bonnelle, C., Bridou, F., Pardo, B. (2002). Modulation of X-ray line intensity emitted by a periodic structure under electron excitation. *Applied Physics Letters*, 81(8), 1524–1526.

Jonnard, P., Wu, M., André, J.-M., le Guen, K., Schäfers, F., Sokolov, A., Filatova, E.O., Verna, A., Wang, Z., Huang, Q. (2018). Dichroism of X-ray fluorescence under standing waves regime in magnetic periodic multilayers [Online]. Available at: https://hal.science/hal-01783440 [Accessed 11 April 2023].

Kao, C.-C., Chen, C.T., Johnson, E.D., Hastings, J.B., Lin, H.J., Ho, G.H., Meigs, G., Brot, J.-M., Hulbert, S.L., Idzerda, Y.U. et al. (1994). Dichroic interference effects in circularly polarized soft-X-ray resonant magnetic scattering. *Physical Review B*, 50(13), 9599–9602.

Kim, S.-K., Lee, K.-S., Kortright, J.B., Shin, S.-C. (2005). Soft X-ray resonant Kerr rotation measurement and simulation of element-resolved and interface-sensitive magnetization reversals in a NiFe/FeMn/Co trilayer structure. *Applied Physics Letters*, 86(10), 102502.

Kuneš, J., Oppeneer, P.M., Mertins, H.-Ch., Schäfers, F., Gaupp, A., Gudat, W., Novák, P. (2001). X-ray Faraday effect at the L2,3 edges of Fe, Co, and Ni: Theory and experiment. *Physical Review B*, 64(17), 174417.

Langer, E. and Daebritz, S. (2010). 75 years of Kossel patterns – past and future. In *Proceedings of the 11th European Workshop of the European-Microbeam-Analysis-Society (EMAS) on Modern Developments and Applications in Microbeam Analysis*, Labar, J.L., Walker, C.T., Zelechower, M., Zieba, P. et al. (eds). IOP Publishing Ltd, Bristol, 33–51.

Le Guen, K., André, J.-M., Wu, M., Ilakovac, V., Delmotte, F., de Rossi, S., Bridou, F., Meltchakov, E., Giglia, A., Nannarone, S. et al. (2019). Kossel effect in periodic multilayers. *Journal of Nanoscience and Nanotechnology*, 19(1), 593–601.

Lee, K.-S., Kim, S.-K., Kortright, J.B. (2003). Atomic-scale depth selectivity of soft X-ray resonant Kerr effect. *Applied Physics Letters*, 83(18), 3764–3766.

McWhan, D.B. (1994). Synchrotron radiation in the study of magnetism. *Journal of Synchrotron Radiation*, 1(1), 83–90.

Mertins, H.-C., Valencia, S., Gudat, W., Oppeneer, P.M., Zaharko, O., Grimmer, H. (2004). Direct observation of local ferromagnetism on carbon in C/Fe multilayers. *Europhysics Letters*, 66(5), 743.

Nefedov, A., Grabis, J., Zabel, H. (2005a). X-ray resonant magnetic scattering on noncollinearly coupled Fe/Cr superlattices. *Physica B: Condensed Matter*, 357(1), 22–26.

Nefedov, A., Grabis, J., Bergmann, A., Radu, F., Zabel, H. (2005b). X-ray resonant magnetic scattering by Fe/Cr superlattices. *Superlattices and Microstructures*, 37(2), 99–106.

Peyrusse, O., Jonnard, P., Le Guen, K., André, J.-M. (2020). X-ray emission from layered media irradiated by an X-ray free-electron laser. *Physical Review A*, 101(1), 013818.

Peyrusse, O., Jonnard, P., André, J.-M. (2021). Maxwell–Bloch modeling of an X-ray-pulse amplification in a one-dimensional photonic crystal. *Physical Review A*, 103(4), 043508.

Purcell, E.M. (1946). Spontaneous emission probabilities at radio frequencies. *Physical Review*, 69(11/12), 681–681.

Raoux, D., Jaouen, N., Tonnerre, J.M., Bontempi, E. (2000). X-ray resonant magnetic scattering: Application to thin films and multilayers. *Acta Physica Polonica A*, 98(5), 483–494.

Röhlsberger, R., Witthoff, E., Gerdau, E., Lüken, E. (1993). Observation of nuclear diffraction from multilayers with a Fe/57Fe superstructure. *Journal of Applied Physics*, 74(3), 1933–1937.

Sacchi, M., Hague, C.F., Pasquali, L., Mirone, A., Mariot, J.-M., Isberg, P., Gullikson, E.M., Underwood, J.H. (1998). Optical constants of ferromagnetic iron via 2p resonant magnetic scattering. *Physical Review Letters*, 81(7), 1521–1524.

Schick, D., Borchert, M., Braenzel, J., Stiel, H., Tümmler, J., Bürgler, D.E., Firsov, A., von Korff Schmising, C., Pfau, B., Eisebitt, S. (2021). Laser-driven resonant magnetic soft-X-ray scattering for probing ultrafast antiferromagnetic and structural dynamics. *Optica*, 8(9), 1237–1242.

Shockley, W. (1939). On the surface states associated with a periodic potential. *Physical Review*, 56(4), 317–323.

Stöhr, J. (1995). X-ray magnetic circular dichroism spectroscopy of transition metal thin films. *Journal of Electron Spectroscopy and Related Phenomena*, 75, 253–272.

Stöhr, J. and Wu, Y. (1994). X-ray magnetic circular dichroism: Basic concepts and theory for 3D transition metal atoms. In *New Directions in Research with Third-Generation Soft X-Ray Synchrotron Radiation Sources*, Schlachter, A.S. and Wuilleumier, F.J. (eds). Springer Netherlands, Dordrecht, 221–250. doi: 10.1007/978-94-011-0868-3_9 [Accessed 11 April 2023].

Strange, P., Hague, C.F., Mariot, J.-M., Durham, P.J., Gyorffy, B.L. (1993). Dichroism in X-ray fluorescence. *Japanese Journal of Applied Physics*, 32(S2), 302.

Tamm, I.E. (1932). On the possible bound states of electrons on a crystal surface. *Physikalische Zeitschrift der Sowjetunion*, 1, 733–735.

Tonnerre, J.M., Sève, L., Raoux, D., Soullié, G., Rodmacq, B., Wolfers, P. (1995). Soft X-ray resonant magnetic scattering from a magnetically coupled Ag $/$Ni multilayer. *Physical Review Letters*, 75(4), 740–743.

Tu, Y., Yuan, Y., Le Guen, K., André, J.-M., Zhu, J., Wang, Z., Bridou, F., Giglia, A., Jonnard, P. (2015). X-ray fluorescence induced by standing waves in the grazing-incidence and grazing-exit modes: Study of the Mg–Co–Zr system. *Journal of Synchrotron Radiation*, 22(6), 1419–1425.

Verna, A., Capelli, R., Pasquali, L. (2021). Resonant soft X-ray reflectivity in the study of magnetic properties of low-dimensional systems. *Magnetochemistry*, 7(10), 136.

Yeh, P., Yariv, A., Cho, A.Y. (1978). Optical surface waves in periodic layered media. *Applied Physics Letters*, 32(2), 104–105.

Zak, J., Moog, E.R., Liu, C., Bader, S.D. (1991). Magneto-optics of multilayers with arbitrary magnetization directions. *Physical Review B*, 43(8), 6423–6429.

Zegenhagen, J. and Kazimirov, A. (2013). *The X-ray Standing Wave Technique: Principles and Applications*. World Scientific, Singapore [Online]. Available at: https://www.worldscientific.com/worldscibooks/10.1142/6666#t=aboutBook.

Zhang, J.-P., Pendenque, C., Le Guen, K., Delaunay, R., Vickridge, I., Schmaus, D., Fu, Q.-G., Jonnard, P. (2019). Kossel interferences of proton-induced X-ray emission lines to study thin film waveguides. *Nuclear Instruments and Methods in Physics Research Section B: Beam Interactions with Materials and Atoms*, 452, 12–20.

7

Applications of Artificial Bragg Structures

Jean-Michel ANDRÉ, Karine LE GUEN and Philippe JONNARD
LCPMR, CNRS, Sorbonne Université, Paris, France

7.1. Wavelength dispersion and monochromatization

7.1.1. *Wavelength dispersion*

The primary characteristic of a Bragg structure in the X-ray radiation domain is its selectivity, in other words, its aptitude to reflect with generally good efficiency only a narrow band of wavelength for a glancing angle greater than the critical value. As soon as fabrication of artificial structures was mastered, this property was put to use to create dispersive elements for soft X-ray radiation spectrometry and monochromators used to select a specific wavelength for a given objective. Prior to this, wavelength dispersion was obtained for long wavelengths (typically greater than 2 nm), using Langmuir–Blodgett mirrors formed of stacked layers composed of saponin molecules of variable length or using diffraction gratings. However, these devices have many limitations: Langmuir–Blodgett structures cannot reflect very long wavelengths, the reflectance is quite low, they are not very stable under irradiation and the gratings only provide noteworthy efficiency at a grazing angle and require a sometimes delicate mechanical assembly. For chemical analysis by X-ray spectrometry, in particular concerning light elements, standard Bragg mirrors have been set up in laboratory instruments (Hombourger et al. 1999; André et al. 2005; Polkonikov et al. 2021) or commercially available devices; among these, we can, for example, mention the Castaing electron microprobe or wavelength-dispersive X-ray fluorescence spectrometers.

The primary disadvantage of multilayer Bragg mirrors in terms of selectivity is their moderate resolving power ($\frac{\Delta\lambda}{\lambda}$ in the order of 10^{-2}) primarily associated with the high photo-absorption in the soft X-ray range. As discussed in Chapter 5, different strategies have been implemented to improve the resolving power. A highly effective but still relatively expensive technique is the use of mirrors etched according to the profile of a lamellar grating (see Figure 3.1a) (Sammar and André 1993; André et al. 2002; Benbalagh et al. 2005; Jonnard et al. 2009; Kozhevnikov et al. 2010). It is possible to reduce the bandwidth by a factor of 5 with an acceptable reduction of the reflectance (see Figure 7.1 and Chapter 5). These mirrors also make it possible to increase the signal-to-noise ratio, and therefore the sensitivity limit, by reducing specular reflection at zero order as shown in Figure 7.1 (Le Guen et al. 2017). In this figure, we see that improvement of the resolution due to etching reduces the K emission of oxygen at order 2 (which is not readily visible with a standard unetched mirror) and the K emission of silicon at order 4. With a reduction in the specular continuous background, the K emission of silicon at order 3 can be resolved. The standard mirror is composed of 300 Mo/B_4C bilayers with a period of d equal to 6 nm, the period D of the grating is 650 nm, and the ratio Γ is 0.54.

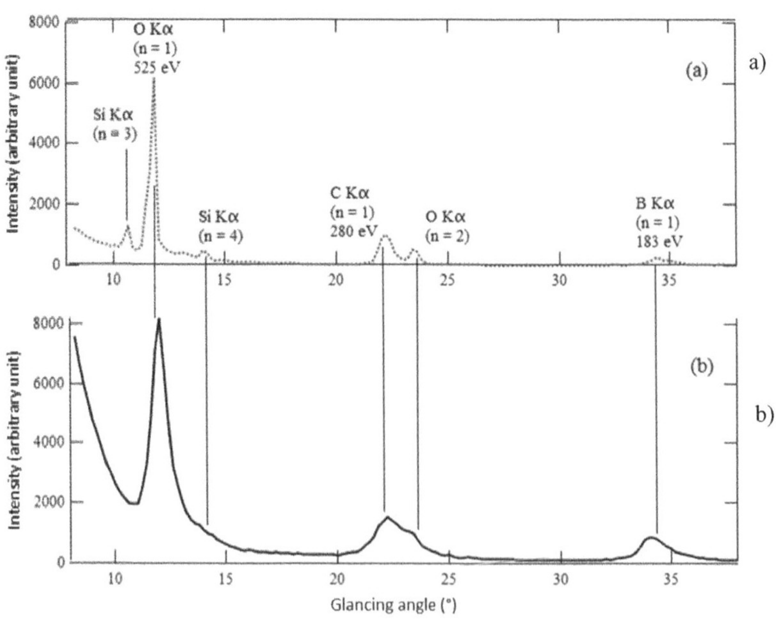

Figure 7.1. *Spectral analysis of a film of boron carbide B_4C deposited on a silicon substrate, using a standard multilayer mirror and an etched mirror; the emission lines are annotated with the name of the emitting element and with the diffraction order (n)*

These mirrors can be cylindrically curved to be inserted into Johann curved crystal spectrometers; thus, modification of the emission bands can be observed as a function of the physical and chemical state of the chemical element with performances close to standard diffraction gratings (Jonnard et al. 2012; Le Guen et al. 2017).

Another type of non-plane crystal spectrometer proposed by L. von Hámos in 1932 (von Hámos 1932), which consists of a cylindrically curved crystal capable of diffracting hard X-ray radiation, has benefitted from artificial Bragg structures and extended to the soft X-ray radiation and EUV domain (Borozdin et al. 2008). This type of highly luminous spectrometer is well suited to low-intensity sources such as energy plasmas of the order of mJ.

A notable contribution of Bragg structures to wavelength dispersion is the creation of diffraction gratings covered with a multilayer deposit that operates as a Bragg reflector. Using this configuration, high efficiency can be obtained for non-grazing angles of incidence, in other words, beyond quasi-total reflection. This type of grating was developed from the 1980s onwards (Keski-Kuha 1984; Jark 1986; Rife et al. 1989; Cruddace et al. 1990) and has led to multiple creations primarily in the form of echelette or blazed gratings intended for use in the spectral domain qualified as soft X-ray radiation, with photon energy between 1 and 5 keV (Voronov et al. 2014; Senf et al. 2016; Lubov and Goray 2019; Sokolov et al. 2019). However, the use of these gratings is limited by the fact that the resonance conditions given by equations [4.42] and [4.43] must be verified for optimal efficiency.

An ingenious technique allowing a very high resolution to be obtained (1 meV around 1 keV) was proposed and implemented by Voronov et al. (2007). It consists of cutting and polishing a multilayer deposited on an echelette grating, offering the incident radiation a high density of grooves[1], as shown in Figure 7.2. Therefore, 27,060 grooves/mm were reported in Prasciolu et al. (2015). The application of this very high angular dispersion and high-resolution grating is envisaged as a dispersive element for performing resonant inelastic X-ray scattering (RIXS). If the radiation to be scattered is in Bragg resonance (in the sense of volume diffraction, see equation [4.36]), then the diffraction efficiency can be very high.

1. The resolving power of a diffraction grating increases as the number of illuminated grooves increases.

Figure 7.2. *Scheme of a periodic multilayer deposited on an echelette grating then cut and polished to offer a high density of grooves. For a color version of this figure, see www.iste.co.uk/andre/xrayradiation.zip*

The performances of this device are remarkable. Therefore, Prasciolu et al. (2015) reported an efficiency of approximately 30% at a wavelength of 13.2 nm for an echelette grating (with period 36.95 nm and blaze angle of 10.5°) covered with a multilayer deposit composed of 3,200 Mo/Si bilayers with a period of 6.7 nm. These diffractive systems are an extension of the technique of truncated multilayers deposited on a flat substrate (Fechtchenko et al. 2002; Bajt et al. 2012) which have the primary disadvantages of a very low dimension and low number of grooves participating in the diffraction, which limits the resolution.

Laue mode volume diffraction (see Chapter 3) is also a way to create dispersive systems offering high efficiency. This approach was selected by Hambach et al. (2001) for the EUV domain. The grating operates by transmission (Laue mode) with a high measured efficiency (close to or greater than 10%) *under Bragg resonance conditions*, at relatively high diffraction orders (from 2 to 5) for a wavelength of 13 nm. This result is in accordance with the coupled-wave theory for the volume diffraction described in Chapter 3 and is an order of magnitude greater than the thin gratings theory. The grating has a high aspect ratio (10:1). It consists of a lamellar structure with a thickness of 700 nm, a period (step) of 200 nm and a period/groove ratio of 0.6 made of a copolymer using a nanofabrication technique. This type of grating is proposed as a condenser in an X-ray microscope with, as a source, a plasma generated by a powerful laser. Therefore, it could be a very useful dispersive element in Castaing microprobes for analysis of light elements with emission lines within the spectral domain covered by this type of grating. In terms of its mode of operation, it has similarities with critical angle transmission (CAT) gratings developed for astrophysics (Heilmann et al. 2022), but without the constraint of working under a grazing angle with a very high groove wall surface quality in terms of roughness.

Polychromatic radiation can be diffracted using a lamellar grating etched in a multilayer Bragg mirror, constituting a device known as a polychromator (Krastev et al. 1996). The process implemented consists of coupling the diffraction, in standard or conical mode[2] from the grating structure as shown in Figure 7.3, to the Bragg diffraction from the multilayer structure. This coupled diffraction, shown in Figure 7.3, can be described from a kinematic point of view by the following equation:

$$sin(\theta_0) = \frac{N(P^2+N^2) \pm \sqrt{N^2(P^2+N^2)^2-(\tilde{P}^2+N^2)[(\tilde{P}^2+N^2)^2-4\,\tilde{P}^2]}}{2\,(\tilde{P}^2+N^2)} \quad [7.1]$$

with:

$$N = n\,\frac{\lambda}{d}, P = p\frac{\lambda}{D}, \tilde{P} = p\frac{\lambda}{D}\,cos(\varphi_0) \quad [7.2]$$

where λ is the wavelength in a vacuum, n is a positive integer corresponding to the Bragg diffraction order and p is an integer corresponding to the grating diffraction order. The geometry and in particular the angles that appear in equations [7.1] and [7.2] are given by Figure 7.3.

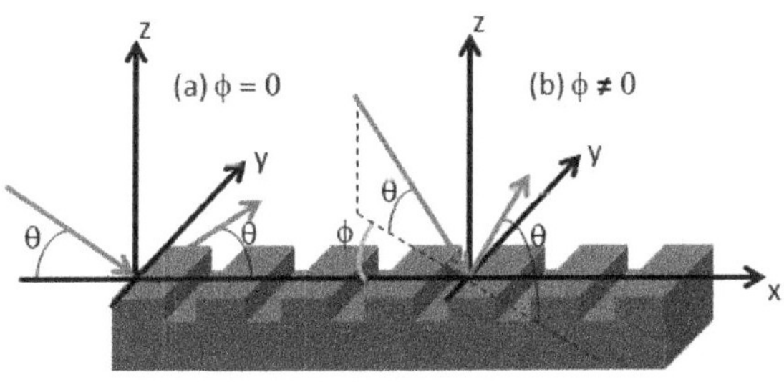

Figure 7.3. *Geometry of diffraction (a) in standard mode and (b) in conical mode*

Equation [6.13] is obtained by combining the generalized Bragg law (equation [7.3a]) and the grating laws in conical diffraction mode (equations [7.3b] and [7.3c]) (see Figure 7.3):

$$sin(\theta_0) + sin(\theta) = N \quad [7.3a]$$

$$cos(\theta)sin(\phi) - cos(\theta_0)\,sin(\phi_0) = 0 \quad [7.3b]$$

$$cos(\theta)cos(\phi) - cos(\theta_0)\,cos(\phi_0) = P \quad [7.3c]$$

2. Diffraction by a grating is said to be conical when the plane containing the incident light ray is not perpendicular to the grating grooves.

Operation in polychromator mode (see Figure 7.4) is explained as follows: if polychromatic radiation reaches this device, it is possible for a wavelength λ_p to satisfy equation [7.1] for a certain value of the integer p. In this diffraction mode, for an incidence (θ_0, ϕ_0), the polychromatic beam is divided spatially into quasi-monochromatic beams centered on the wavelength λ_p given by:

$$\lambda_p = \frac{\left(\frac{n}{d}\right)sin(\theta_0)-\left(\frac{p}{D}\right)cos(\theta_0)cos(\phi_0)}{\left(\frac{n}{d}\right)^2+\left(\frac{p}{D}\right)^2} \qquad [7.4]$$

and whose direction (θ, ϕ) is given by equations [7.3a–c].

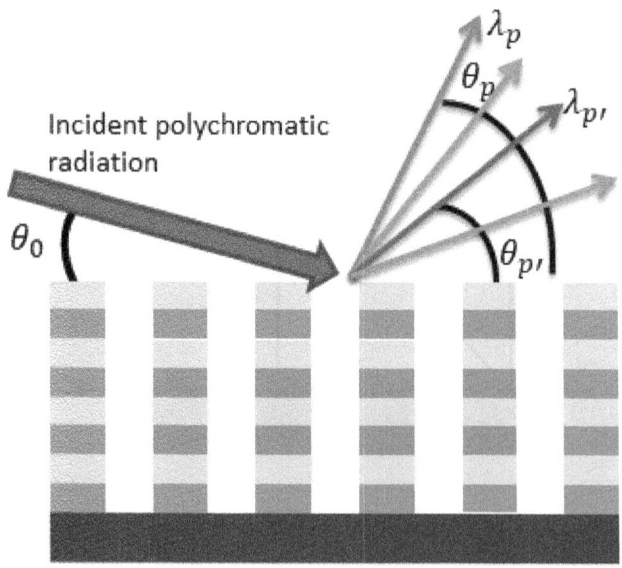

Figure 7.4. *Scheme of a polychromator device obtained by etching a standard Bragg mirror in the direction of depth; the incident polychromatic beam is divided into quasi-monochromatic beams centered spectrally on a wavelength λ_p given by equation [7.4]. For a color version of this figure, see www.iste.co.uk/andre/xrayradiation.zip*

This polychromator can be considered to be a spectral sampler with step $\delta\lambda$ in terms of wavelength, given approximately for an incidence (θ_0, ϕ_0) by:

$$\delta\lambda = \frac{2\,d^2\,cos(\theta_0)cos(\phi_0)}{n^2\,D} \qquad [7.5]$$

This diffraction mode, in standard mode and in conical mode, applied to the emission from a Coolidge X-ray tube (see Chapter 1), is shown in Figure 7.5. This device is useful as a beam separator or spectral sampler for pulse sources because it does not require mechanical displacement of the optical device.

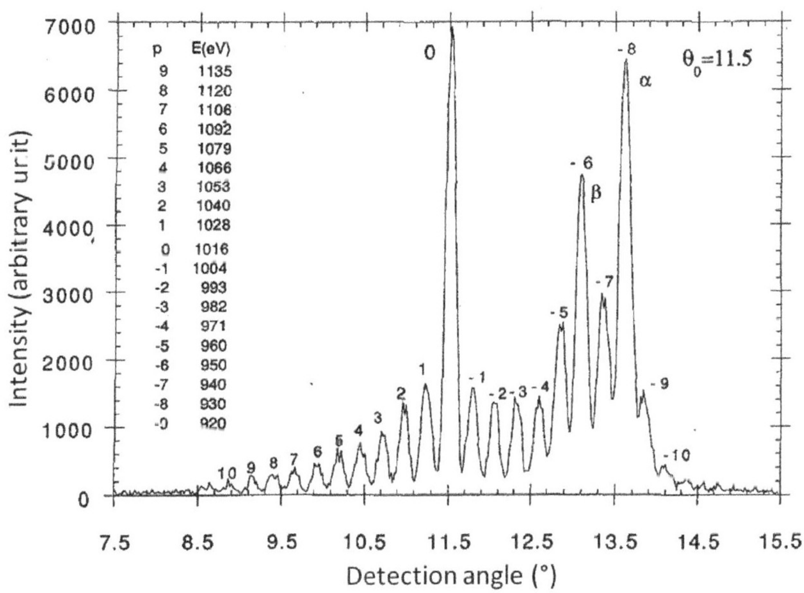

Figure 7.5. *Intensity as a function of the angle of detection θ of the radiation emitted by a copper anode Coolidge tube and diffracted by a lamellar grating etched in a multilayer, operating in polychromator mode in standard diffraction, according to Barchewitz and André (1996)*[3]

Another type of dispersive element was proposed by André (1998). It relies on the use of a semi-transparent multilayer Bragg structure. Dispersion is by transmission and not by reflection. The advantage of this device is its very simple geometric arrangement, which may require no mechanism if a position-sensitive detector such as a charge-coupled device (CCD) camera is used (see Figure 7.6(a)). A very compact device can be created, and the sample to analyze can be very close to the dispersive element, even though, to obtain significant angular scattering, it is still necessary to

3. See doi: 10.1051/jp4:1996474.

move the latter away from the detector. The incident radiation can be a beam of X-rays (X-ray fluorescence) or charged particles: electrons or protons for proton-induced X-ray emission (PIXE), as shown in Figure 7.6(b) (Jonnard et al. 2018).

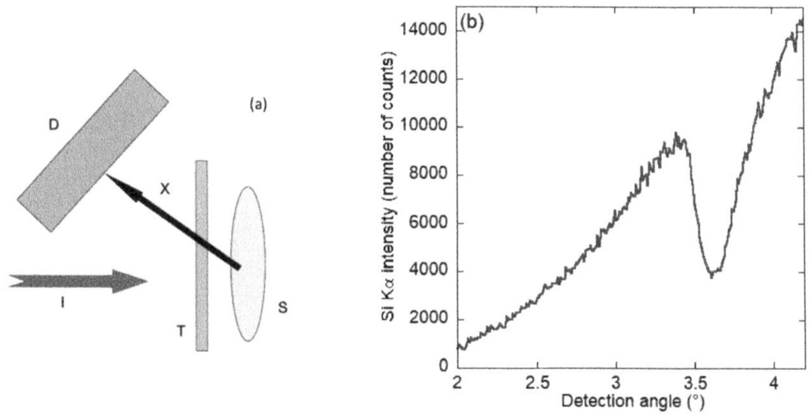

Figure 7.6. *(a) Scheme of the transmission spectrometer: I, incident beam; T, dispersive element (semi-transparent multilayer); D, location detector; X, X-ray radiation emitted by the sample; S, sample; (b) Si Kα spectrum of a silicon substrate obtained with a transmission spectrometer consisting of a multilayer formed of 20 bilayers $B_4C/Pd/B_4C/Y$ with a period of 6 nm. For a color version of this figure, see www.iste.co.uk/andre/xrayradiation.zip*

7.1.2. *Combined monochromatization*

We distinguish between monochromatization and wavelength dispersion by the fact that a monochromator is used for selecting a wavelength or a narrow band of wavelengths while the dispersed spectrum is recorded or analyzed. A wavelength-dispersive device is more of a spectrometry tool, whereas a monochromator is an optical device often providing the additional functions of focusing or collimation, constituting what we can call combined monochromatization. Focusing and collimation with a large numerical aperture (NA) are generally performed with curved optics: focusing using elliptical curvature (see Figure 7.7(a)) and collimation using parabolic curvature (see Figure 7.7(b)).

The possibility of creating multilayer systems with a horizontal period gradient which allow local adjustment of the Bragg angle at the wavelength being selected offers a considerable advantage for the use of curved substrates. The parabolic mirror is used in the powder method (Misture 2008; Shimizu and Omote 2008) and in diffractometry with an assembly referred to as a Göbel mirror (Schuster and Gobel 1995). In the 1950s, M. Montel designed a reflector system for X-rays allowing for two-dimensional focusing or collimation, consisting of two mirrors arranged perpendicularly side by side (Montel 1957). This system is now set up with a multilayer deposit with a horizontal period gradient and has a hybrid version combining an elliptically curved mirror and another curved parabolically allowing a rectangular beam to be formed (see Figure 7.8). The Montel arrangement is more compact and offers an aperture four times larger than the classic Kirkpatrick–Baez scheme.

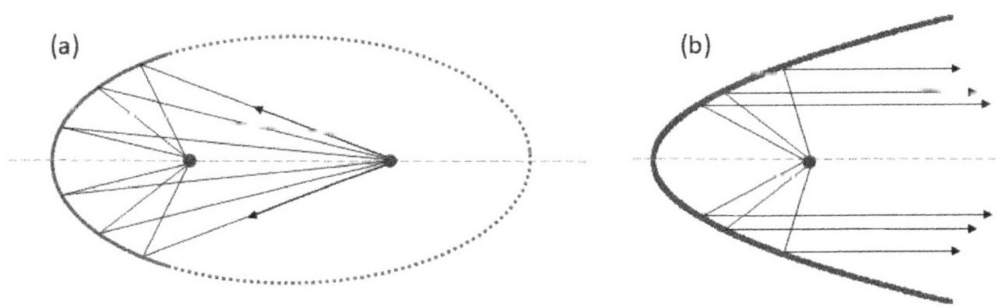

Figure 7.7. *(a) Focusing using an elliptical mirror; (b) collimation using a parabolic mirror*

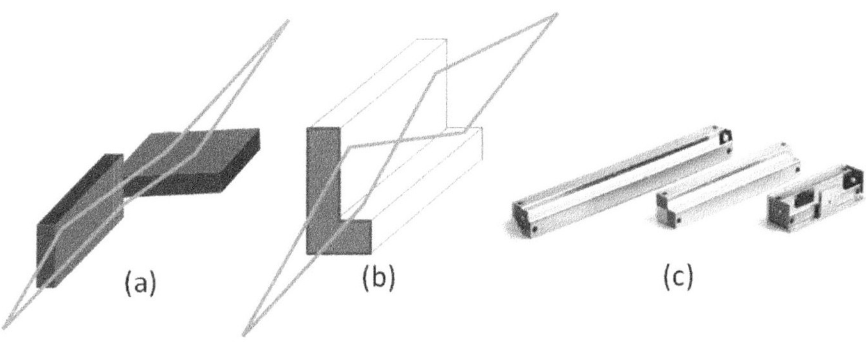

Figure 7.8. *(a) Classic Kirkpatrick–Baez focusing device; (b) Montel focusing device; (c) photograph of Montel systems taken by the company Incoatec (courtesy of that company)*

Combined monochromator systems are also implemented when the divergence of the source of radiation is large and the instrument using this diverging beam (imager, detector, etc.) is too small to intercept the entire beam. An example of this type of situation is given by the beamlines from third-generation synchrotron sources such as the Advanced Photon Source (APS) where radiography and tomography stations are installed. A combined monochromator performing sagittal focusing and which can produce a spatially wide and high-brilliance X-ray radiation beam has been developed for this type of beamline (Wang et al. 2007). The scheme of this device is presented in Figure 7.9. It allows dynamic adjustment of the sagittal radius of curvature but suffers from significant aberrations in the far field, which limits its use to the near field (radiography, tomography).

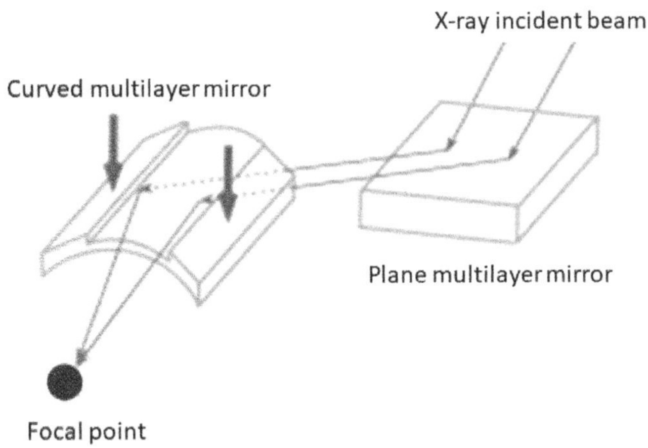

Figure 7.9. *Scheme of a combined monochromator based on artificial Bragg structures allowing sagittal focusing*

COMMENT ON FIGURE 7.9. – *The incident beam, wide and diverging, is reflected by a plane multilayer mirror, then by a curved multilayer mirror, under the action of forces indicated by the blue arrows, and finally is focused. According to Wang et al. (2007)*[4]*, adapted with the permission of the International Union of Crystallography.*

4. See doi: 10.1107/S0909049506050205.

7.2. Focusing using lenses

7.2.1. *Laue lens*

When the beam is not too wide and diverging, for example, in a synchrotron beamline at a moderate distance from the point of emission, focusing can be performed using Fresnel lenses operating by transmission, therefore in Laue geometry. Since their first use in the 2000s[5], analog devices that instead use volume diffraction where Bragg resonance is involved have been considerably developed thanks to nano- and micro-fabrication technologies. We recall that in Laue geometry, Fresnel lenses consist of a series of concentric circular zones with radii r_n that vary with the wavelength λ according to the relationship:

$$r_n(z) = \sqrt{n\,\lambda\, f + \left(\frac{n\,\lambda}{2}\right)^2}\left(1 - \frac{z}{2f + \frac{n\,\lambda}{2}}\right) \qquad [7.6]$$

The integer n which is the index of the radius r_n numbers this radius, $n = 0$ corresponding to the central zone. λ is the wavelength of the radiation, f denotes the focal distance, and z is the depth counted from the external surface of the lens exposed to the incident radiation. The term in brackets in relationship [7.6] translates the Bragg condition.

The characteristics and manufacturing techniques of these lenses differ notably depending on the type of radiation being focused. In the soft X-ray radiation domain, absorption is high and the opaque zones of the lens require a thinner material thickness than for hard X-ray radiation: to obtain a lens with high efficiency, the thickness of the lens must be greater than the diffraction extinction distance, so several microns in the hard X-ray radiation domain. Lenses of this kind are qualified as high aspect ratio structures. In the field of soft X-ray radiation requiring a relatively low aspect ratio, fabrication can make use of an electron beam lithography technique. This can no longer be used in the hard X-ray radiation domain, and in this case, two other techniques have been applied, namely rolling and the multilayer Laue lens (MLL). The technique of rolling consists of alternating between depositing layers of materials with index contrast Δn on a cylindrical core and slicing the structure obtained according to the desired thickness, perpendicular to the axis of the cylinder. The optical thickness t_{op}, which corresponds to a phase shift of π, is given by:

$$t_{op} = \frac{\lambda}{2\Delta n} \qquad [7.7]$$

5. Fresnel lenses for X-ray radiation but not using Bragg structures were designed and developed in the first half of the 20th century, implementing highly absorbing materials (Baez 1952; Kirz 1974).

Very high values of aspect ratio can be obtained using this technique. Therefore, an aspect ratio of 792 with energy of 20 keV, giving a lens thickness of 32 microns and allowing efficiency of 27% to be achieved at first order, was obtained by Koyama et al. (2012). However, the method of deposition, generally the technique of atomic layer deposition, very often introduces stacking errors which are expressed as optical aberrations, sources of deterioration of the resolution. For applications in microscopy, the linear core size must be below 10 nm, which is difficult to access using the rolling technique. In addition, defects inherent to the cylindrical core (roughness, radial inhomogeneity) and those introduced by slicing are the source of problems affecting the spatial resolution.

The alternative approach uses linear MLLs based on a deposit according to the law given by equation [7.6] of alternating layers with a high absorption contrast on a super-polished plane substrate. The MLLs are divided into different types, planar, inclined, cornered and curved, which are represented in Figure 7.10. They can be looked at as the degenerated linear version of circular lenses.

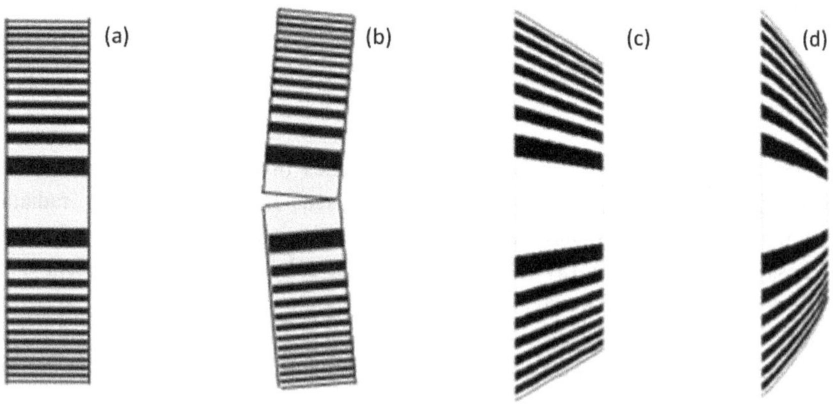

Figure 7.10. *Different types of MLL: (a) flat; (b) tilted; (c) wedged; (d) curved. According to Yan et al. (2007[6], 2007), with the permission of the American Physical Society*

Flat MLLs are the simplest to create but their resolution and efficiency remain limited following dynamic volume diffraction effects (Yan et al. 2007): the profile of the layers, in other words, changes in the radius as thickness z varies, does not

6. See doi: 10.1103/PhysRevB.76.115438.

follow equation [7.6]. To limit these effects, we can tilt two MLL halves with respect to the optical axis in order to satisfy the Bragg law with the most external zones, as represented in Figure 7.10(b). However, this technique remains insufficient for efficiencies greater than 50% and core sizes of the order of a nanometer.

To obtain a high resolution, which is important in microscopy for example, lenses must be made with a wide NA, a quantity defined by the relationship:

$$NA = sin(\theta_{\mathrm{NA}}) \qquad [7.8]$$

where θ_{NA} is equal to half of the angle formed by the most deflected light ray and the least deflected light ray.

However, it is necessary to adjust the targeted NA at the angular width Ω of the diffraction pattern of the diffracting structure. If the targeted NA is greater than Ω, only a restricted part of the lens will be useful, which will reduce the resolution. To make the entire lens effective, it is necessary for all zones to be in Bragg resonance conditions, which implies a suitable orientation of the layers with regard to the incident radiation. To do this, it is necessary to profile the zones according to the optical axis to satisfy *the Bragg resonance locally*, which is the objective of the wedged MLL represented in Figure 7.10(c). Figure 7.11, taken from Murray et al. (2019), gives details of the operation of a wedged MLL: efficiencies close to 62% have been reported for this type of linear lens with a core size of 13.1 nm (Koyama et al. 2011a), better performances than those obtained with the rolling method reported by the same group (Koyama et al. 2011b).

Linear MLL can be associated in a crossed mode to obtain rectangular cores with a surface of the order of 50 nm^2 for a photon energy between 10 and 20 keV (Koyama et al. 2011a; Bajt et al. 2018). MLLs with profiled zones, according to confocal parabolas in the case of a planar incident wave or ellipses in the case of a spherical incident wave, should perform even better (Yan et al. 2007), but they remain very difficult to create.

MLLs are promising optical elements in the X-ray radiation domain, in particular in the domain of high-resolution microscopy. Their performances have been tested at 17.3 and 34.5 keV (Murray et al. 2019). They have the potential to provide beams focused at nanometric scale: a core dimension of 5 nm or less can be envisaged in the short term. Lenses of this type have already been used in X-ray fluorescence scanning microscopes, which produce a cartography of elements in nanoparticles (Kang et al. 2013).

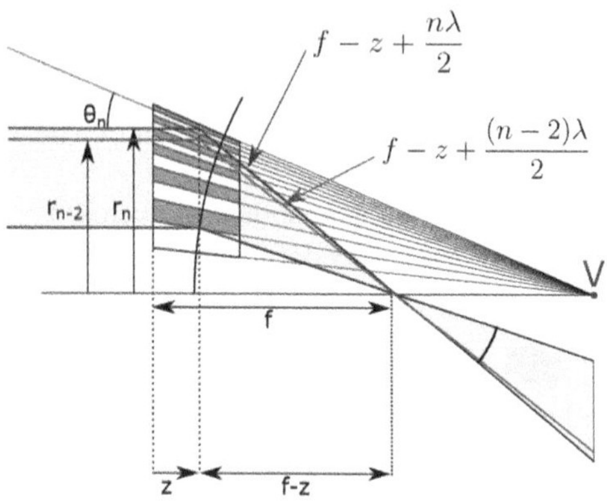

Figure 7.11. *Scheme and principle of a wedged MLL. For a color version of this figure, see www.iste.co.uk/andre/xrayradiation.zip*

COMMENT ON FIGURE 7.11.– *The layers are tilted with respect to the external surfaces and oriented on a circle centered on the summit V. The blue radius is deviated at the zone with radius r_n and the red at the zone with radius r_{n-2} corresponding to a step difference equal to $\frac{\lambda}{2}$; these two radii converge at the core of the lens located at the distance f on the optical axis **Z**. The numerical aperture associated with the two extreme rays in blue is indicated; beyond this core, the beam diverges with an angle equal to 2NA (Murray et al. 2019)[7]. © The Optical Society.*

7.2.2. Bragg–Fresnel lens

This type of lens was initially developed by a group of researchers at the Chernogovolka microelectronics institute in Russia (Aristov et al. 1986). Its principle is described in Chapter 4. Its primary advantage lies in the reduction of chromatic aberrations compared to systems that focus using grazing incidence mirrors. The main application of these lenses is the creation of microprobes intended for microanalysis of materials containing emitting elements in the soft X-ray range, in other words, light elements. This application requires two-dimensional focusing. This is obtained by two crossed linear lenses or an elliptical lens (Erko 1994). X-ray microfluorescence using synchrotron radiation benefitted from this with harder radiation (12 keV) and beam sections of the order of a square micron (Mirone et al. 1994). Using off-axis geometry, the contrast can be increased by eliminating the zero

7. See doi: 10.1364/OE.27.007120.

order of focal zones (Idir et al. 1997), as shown in Figure 7.12. Such off-axis lenses are obtained by etching only the end part of the virtual lens centered with a high number of grooves in order to maintain a large angular aperture. They can have either linear or elliptical geometry and be assimilated to lamellar gratings with a variable step (Harada 1990). This type of off-axis optical device has been implemented, among other things, to perform plasma imaging (Cauchon et al. 1998) and beam control in synchrotrons.

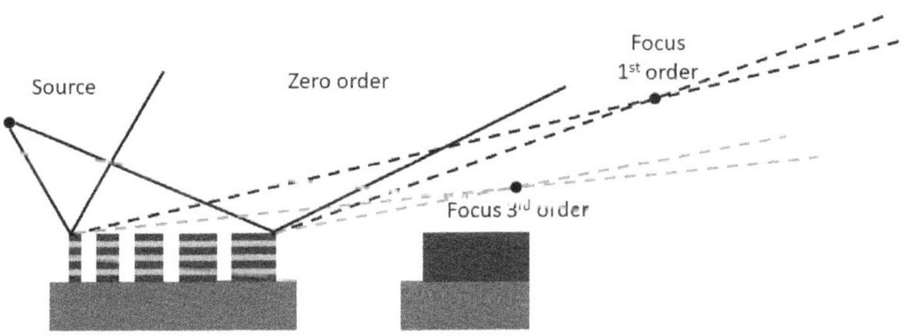

Figure 7.12. *Schematic diagram of a Bragg–Fresnel lens operating in off-axis mode*

7.3. Polarimetry, interferometry and metrology of X-ray radiation

By metrology of radiation, we mean measurement of radiation characteristics or of the beam carrying it: the nature and parameters of the polarization of the radiation, of the degree of coherence, of the quality of the wavefront, etc.

7.3.1. *Polarization*

The nature of the polarization (linear, circular, elliptical, natural, in other words not polarized) of the X-ray radiation depends significantly on the type of source:

– a Coolidge tube essentially emits non-polarized radiation;

– a curving magnet on a synchrotron emits linearly polarized radiation in the orbital plane with a degree of polarization that increases, while moving away from the plane;

– a helicoidal undulator emits radiation of which the polarization can be linear, elliptical or circular depending on its configuration.

The characteristics of polarization of the radiation (Stokes parameters, Müller matrices, DOP, etc.[8]) are important for certain applications (optical and magnetic measurements in particular) and can differ notably from the characteristics of the source following the influence of optical elements on the beamline. A polarimeter designed to obtain the four Stokes parameters is based on a phase retarder combined with an analyzer. In the VUV domain, retarder and analyzers are based on reflective optics. Therefore, VUV polarimetry was first performed in 1986: it used two multilayer Bragg mirrors working under a 45° incidence, and it was tested on the VEPP-2M synchrotron with a curved magnet source and a helicoidal undulator (Gluskin et al. 1986).

In the soft X-radiation domain, combined transmission and reflection Bragg systems are mostly used: the phase shift Δ between the component *s* (*TE*) and the component *p* (*TM*) induced by a multilayer Mo/Si has been calculated (Kortright and Underwood 1990) and it became apparent that a significant phase shift between these two components could be obtained using a transmission plate operating under Bragg conditions near an absorption threshold of an element making up the structure, as shown in Figure 7.13. Considerable development of these retarding plates and polarimeters then ensued using these devices (Kortright et al. 1992; Schäfers et al. 1999; MacDonald et al. 2008).

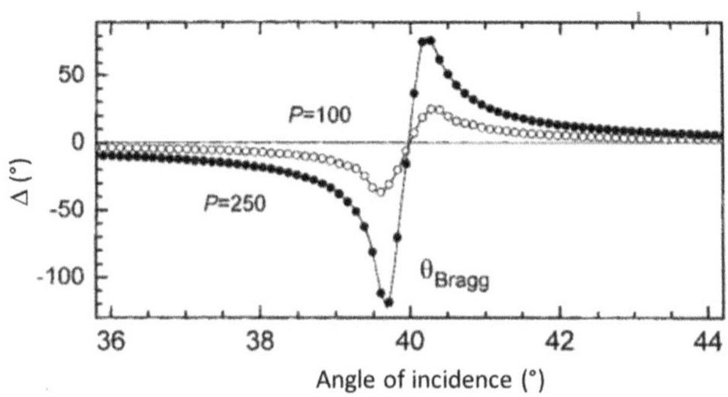

Figure 7.13. *Phase retardance Δ as a function of the angle of incidence obtained with a transmission retarding plate, made up of a periodic multilayer Cr/Sc with period of 2.21 nm with various numbers of bilayers (100, 250) for a photon energy of 397 eV (near the threshold Sc $L_{2,3}$); reproduced with the permission of Schäfers et al. (1999)[9]. © The Optical Society*

8. The characteristics of polarization of radiation are created using different tools (Stokes parameters, DOP, etc.), see Appendix 2.
9. See doi: 10.1364/AO.38.004074.

The transmission retarding plates can also be used as polarizers, capable of transforming linearly polarized radiation to circularly polarized radiation and vice versa (Lang and Srajer 1995).

In the EUV domain, absorption of the radiation becomes prohibitive for the use of transmission plates; it was proposed to use linearly chirped mirrors as a phase retarder: with an Mo/Si mirror with a period centered on 14.7 nm, phase shifts from 5.95° to 245.10° at a photon energy of 90 eV were obtained when the number of bilayers increased from 2 to 20 (Yang et al. 2021). In addition, multilayer Bragg mirrors with horizontal period gradient operating near the Brewster angle (glancing angle of approximately 45°) can constitute tunable polarizers by simple translation with respect to the incident beam (Kortright et al. 1995).

7.3.2. *Coherence*

The primary contribution of artificial Bragg structures to the study of coherence in the X-ray radiation domain was made by means of interferometry. Interferometry by division of amplitude benefitted from the technology of so-called semi-transparent separating plates developed at the end of the 1980s (Hawryluk et al. 1987; Susini et al. 1988; Khan Malek et al. 1989). These plates most often consist of a multilayer structure deposited on a membrane generally made of silicon nitride (Sobierajski et al. 2013), which generates mechanical stress problems.

However, technologies have been developed that enable self-supported structures to be created (Haga et al. 1998). Interferograms were recorded using a Michelson interferometer equipped with a separating plate with an Mo/Si multilayer deposit (Delmotte et al. 2002) and using a Mach–Zehnder interferometer (Libby et al. 1996). The longitudinal coherence and the emission of a collisional X-ray laser at 15.5 nm was measured using a Mach–Zehnder interferometer (Celliers et al. 1995; Da Silva et al. 1995) and at 14.68 nm using a Michelson interferometer, which were equipped with separating plates with multilayer deposits (Smith et al. 2003). Using these instruments, measurement of the electronic density within the emissive plasma of these lasers has also been possible.

7.4. Instruments and imaging at quasi-normal incidence

Instruments using artificial Bragg mirrors operating at quasi-normal incidence are at the center of very important developments from an economic and financial point of view. From this point of view, lithography using radiation at the wavelength of 13.5 nm (EUVL) combines these two aspects: economic because EUVL should be the next technology used to fabricate semiconductor integrated circuits and financial because the amounts and resources invested in this technology are considerable. We estimate this to be approximately 6 billion euros, in 17 years of research, the budget the Dutch company ASML would have required to develop the first series of lithography machines EUVL suitable for commercialization. From a financial point of view, we also find the latest-generation space telescopes developed for astrophysics exploring the EUV domain and soft X-ray radiation.

From the instrumental optics point of view, instruments with quasi-normal incidence, compared to the optical devices operating under grazing incidence, offer the advantages of being compact in nature (important for space telescopes) and a larger NA (including improved spatial resolution[10]), leading to an improved image quality. The multilayer deposits are made on curved surfaces (sphere, parabola or ellipse) with the possibility of varying the period horizontally to match it to the angle of incidence to satisfy the Bragg condition. We discuss the three main applications of these optics, in other words, the imaging for EUV lithography, telescopes intended for astrophysics operating in the EUV and soft X-ray domains, and EUV microscopy.

7.4.1. *EUV lithography*

In order to continue miniaturization of integrated electronic components such as microprocessors, it is necessary to reduce the size of the printed patterns on the semiconductor. The technology used is photolithography by projection[11]. Yet, the minimum size of the grooves is limited by the diffraction effect, therefore by the radiation wavelength used (Rayleigh criteria). This latter is currently 193 nm, and we aim to use wavelength radiation close to 13.5 nm. This value results from a compromise between multiple demands: availability of quite a powerful radiation source, reflective optics that are quite effective in terms of reflectance and beam quality, the sensitivity of the resin, etc.

After years of studies that began in the 2000s, involving a large number of public and private research centers in several countries (United States of America, European countries, China and Russia), different prototypes of an EUVL machine

10. Diffraction-limited resolution R_{dl} is given by $R_{dl} = \frac{0{,}44\,\lambda}{NA}$, where λ is the wavelength of the radiation and NA is the numerical aperture.
11. Short-wavelength electromagnetic radiation traverses the drawing of the circuit to be printed on photosensitive resin. The analogy between the resin and stone (*lithos* in Ancient Greek) that is engraved to write a message explains the terminology "lithography".

have been created. To date, the company ASML, in collaboration with the German companies Zeiss and Trump, has constructed a machine for which the principle is presented in Figure 7.14 that it is preparing to supply to the main funders, producers of these electronic chips: Intel, Micron, TSMC, Samsung, SK Hynix, etc. It comprises a source of radiation at 13.5 nm composed of a plasma generated by interaction of a laser beam with droplets of tin, optical assemblies containing Mo/Si multilayer Bragg reflector mirrors. From an optical point of view, as shown in Figure 7.14, a typical machine involves a system named *Illuminator Optics* whose function is to carry the radiation from the source towards the reflective mask, a projection device (6-Mirror Projection Optics) that projects the pattern drawn on the mask to the surface of the wafer. Significant research has been performed on the multilayer deposits based on the Mo/Si couple, and prior to use, they offer reflectance close to 70% at the wavelength of use.

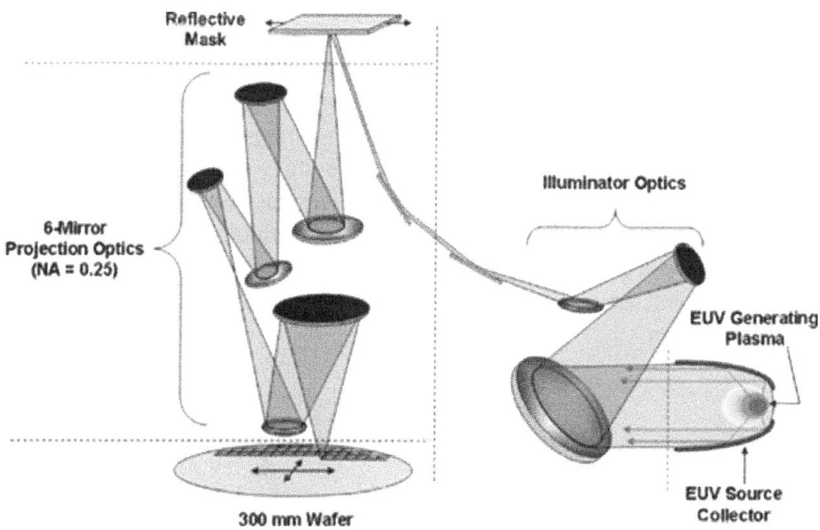

Figure 7.14. *Schematic diagram of the EUV lithography system. Multilayer Bragg mirrors are involved in the "Illuminator Optics" optical device and in the "Projection Optics" projection system which offers a numerical aperture (NA) of 0.25, according to Wurm and Kemp (2006)*

Currently, research is ongoing to increase the NA while maintaining the same wavelength (13.5 nm) in a program called *High NA technology* where the NA value of 0.5 is targeted. There is also research aiming to use shorter wavelengths of the order of 6.7 nm with multilayer mirrors adapted in the context of a project named *Beyond Extreme Ultraviolet Lithography* (Uzoma et al. 2021). The principles and technologies used – in particular those concerning the radiation source, optics, resin, problems, performances and prospects – are explored in several specialized works (Bakshi 2018; Erdmann 2021).

7.4.2. *Telescopes for EUV astrophysics*

This domain benefitted very early on from progress in the field of multilayer deposits at nanometric scale. Several missions have been carried out to study the Sun, in particular its corona, using telescopes with two mirrors (Cassegrain type) where the Bragg structure is divided into several zones for observation within a specific spectral band. Therefore, the Extreme Ultraviolet Imaging Telescope, installed on the Solar Heliospheric Observatory (SOHO, the result of a collaboration between NASA and ESA) mission, uses four quadrants on the primary and secondary mirror of the Cassegrain telescope (see Figure 7.15(a)), tuned on the bands centered on the wavelengths 30.4, 17.1, 19.5 and 28.4 nm corresponding respectively to the emissions He II, Fe IX-X, Fe XII and Fe XV. A rotating obturator means that only one quadrant at a time can be operated for the study of a given spectral band (Delaboudinière et al. 1995). The multilayer mirrors are based on the Mo/Si couple. The instrument SUIVI equipped with a generalized Cassegrain telescope (Martínez-Galarce et al. 2013) operates on the same multiband principle but with six spectral bands centered respectively on the wavelengths 9.39, 13.12, 17.11, 19.51, 28.42 and 30.38 nm.

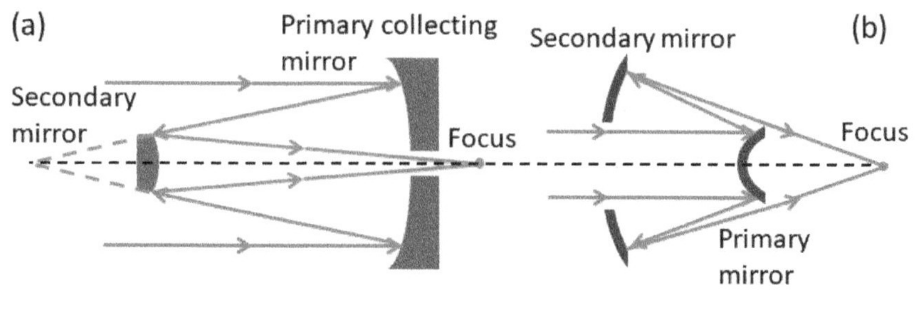

Figure 7.15. *Schemes of (a) a Cassegrain telescope and (b) a Schwarzschild objective*

Observation of the Sun using telescopes equipped with mirrors covered in periodic multilayers operating as Bragg reflectors continues; a recent project in the context of the Solar Orbiter mission, aiming to explore the internal heliosphere using an instrument named the Extreme Ultraviolet Imager (EUI), is being developed in a collaboration between NASA and ESA (Rochus et al. 2020). It will comprise three telescopes and two high-resolution imagers. Two wavelength bands will be analyzed – one at 17.4 nm and the other at 30.4 nm using a $[SiC/Mo/Al]_4/Al/SiC/Mo/Al]_{30}$ composite bi-band Bragg structure. The first multilayer is composed of 30 SiC/Mo/Al trilayers and is designed to reflect radiation with wavelength 17.4 nm, whereas the second with four trilayers of the same composition is designed for the band at 30.4 nm (Hecquet et al. 2009). The layer of aluminum between the two Bragg structures has the role of a buffer, and its thickness has been optimized to reject undesirable wavelengths. These aluminum-based multilayer optics have been studied and developed at the Laboratoire Charles Fabry at the Institut d'optique in Palaiseau, France (Meltchakov et al. 2009, 2011), as was the case for Mo/Si multilayers of the EIT instrument.

7.4.3. *X-ray/EUV microscope*

The Schwarzschild objective, a system with two mirrors (see Figure 7.15(b)), is a device selected as a priority for creating a microscope that can be used to obtain a magnified image of a micro object with a high spatial resolution (submicronic). It presents the advantage, compared to a single-mirror system (playing the role of condenser) combined with a Fresnel lens (playing the role of lens), of having no astigmatism, spherical aberrations or comas, while offering a wide field.

Microscopes with Schwarzschild objectives were developed from the 1990s, most often with a laser as a source in the EUV domain generated by a plasma. DiCicco et al. (1992) thus created a device of this type operating at the wavelength 18.2 nm with mirrors operating at quasi-normal incidence with an Mo/Si multilayer Bragg structure.

A Schwarzschild optic can itself be used as a condenser in association with a Fresnel lens playing the role of lens according to the scheme presented in Figure 7.16, to create a microscope with resolution close to the wavelength of the source (capillary discharge laser), so 46.9 nm (Brewer et al. 2008). The condenser is made up of two spherical mirrors with an Sc/Si multilayer deposit offering overall reflectance of 13%.

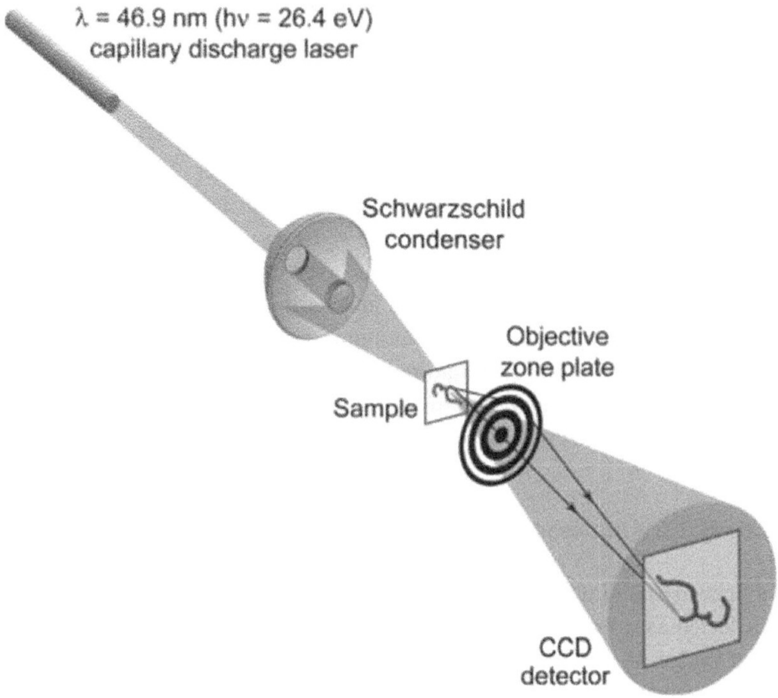

Figure 7.16. *Scheme of an EUV Schwarzschild condenser microscope and Fresnel lens objective, according to Brewer et al. (2008)*[12]. *© The Optical Society*

7.5. Instruments and imagery under grazing incidence

Instruments incorporating Bragg mirrors and operating under grazing incidence are primarily used in astrophysics and for third-generation synchrotron beamlines, in the very hard X-ray radiation domain, typically for energy photons greater than 10 keV. For hard radiation up to 10 keV, Wolter I[13] telescopes with parabolic and hyperbolic curved mirrors are used (see Figure 7.17). The surfaces of these super-polished mirrors are covered with a single layer of film made of a heavy material, generally iridium or gold.

12. See doi: 10.1364/OL.33.000518.

13. In 1952, Hans Wolter designed optics initially intended for use in an X-ray microscope. He proposed three arrangements (Wolter I, II, III) that combine two types of confocal mirrors with circular symmetry (parabolic, hyperbolic and elliptical) operating under grazing incidence. Type I (hyperbolic + parabolic) was then selected to create space telescopes in a configuration of several mirrors arranged in nested shells, mainly due to the shorter focal length of this type compared to the two other types.

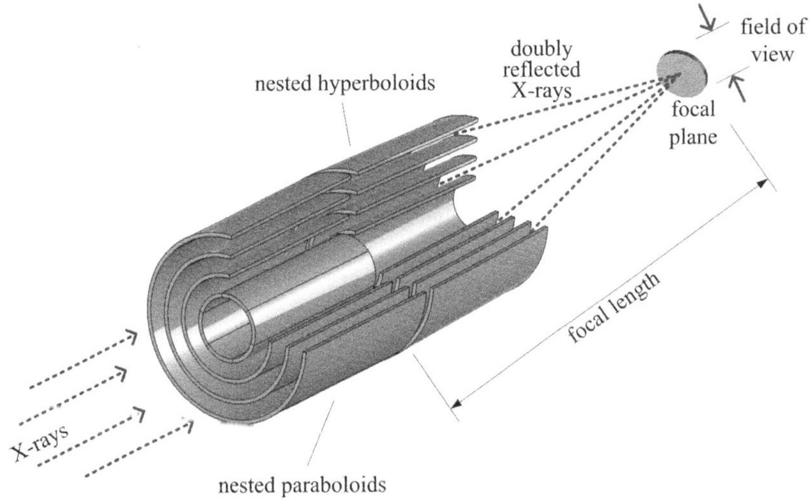

Figure 7.17. *Principle of a type I Wolter telescope; the series of mirrors collecting the X-ray radiation are paraboloids and the series of mirrors focusing the radiation of the hyperboloids are doubly reflected rays. According to Döhring et al. (2015)*[14]

The first focusing extrasolar X-ray telescope, named Einstein Observatory, operating on the basis of this principle was launched in 1978. It was followed by a series of similar assignments (Chandra by NASA, XMM-Newton by the ESA, etc.). When multilayer deposits at nanometric scale became available, it was then possible, applying the principle of super-mirrors (see Chapter 4), to extend the observable spectrum to high energies, in other words, for photons up to 100 keV. Therefore, NASA's recent mission, known as the Nuclear Spectroscopic Telescope Array (NuSTAR), uses multilayer structures Pt/C and W/Si with a period gradient created at the Danish Technical University in Denmark (Harrison et al. 2013). This approach also seems useful for lower energies (0.1–10 keV) using couples of iridium- and carbide-based materials: Ir/B_4C and Ir/SiC, as indicated in the studies for the Advanced Telescope for High-Energy Astrophysics (ATHENA) mission (Jakobsen et al. 2011). We also note that for the Chinese satellite X-ray Timing and Polarization (XTP), a space telescope has been developed, based on the same principles and operating in the 0.5 to 10 keV domain (Wang et al. 2014).

Super-mirrors capable of creating compact beamlines for synchrotron sources emitting in the hard X-ray radiation domain have been used in all third-generation synchrotrons. They can reflect across a wide spectral band at glancing angles that

14. See doi: 10.1117/12.2191112.

are two or three times higher than the critical angle of mirrors operating in the total reflection regime of a homogeneous environment, such that the length of the optic and the value of the radius of curvature are reduced in the inverse ratio of the angle, by a factor of 2 to 3. In Chapter 4, we saw that very complex and powerful algorithms are used to create these super-mirrors. But in many cases, the first generations of these devices were created according to a variation law $d_j = a/(b+j)^c$, in which *j* is the number of the layer, where the coefficients *a, b, c* and the number of layers must be optimized by minimizing a merit function (Joensen et al. 1995; Windt et al. 2000).

7.6. Elements for X-ray radiation sources

7.6.1. *Introduction*

Artificial Bragg structures are involved at two levels. They can be implemented as either:

– optical devices inside a source to perform different functions: amplify (optical cavity), focus or collimate, etc., the radiation;

– elements that generate radiation (radiators) by using different phenomena: quasi-Cherenkov emission (transition or parametric radiation) or stimulated emission; in the latter case, resonance effects related to the periodicity allow amplification, filtering and monochromatization of the radiation.

7.6.2. *Optical elements for sources of X-ray radiation*

The appearance in the 1980s of interferential multilayer structures operating with a high reflectance under quasi-normal incidence practically coincided with the development of plasma X-ray lasers[15] generated by powerful lasers and using collision or recombination processes (Matthews et al. 1985; Giulietti and Gizzi 1998). This situation allowed optical cavities to be created for these lasers, enabling amplification of the stimulated radiation according to the classic scheme already operating in the long wavelength domain. Initial work used a semi-open cavity, which provided only one open passage for the radiation in the emissive plasma (Suckewer et al. 1985), and then different schemes of semi-closed cavity were proposed (Ceglio et al. 1986), as shown in Figure 7.18: the cavity includes a device

15. The term X-ray laser is used in this chapter without specifying the spectral domain concerned (EUV, soft or hard X-ray radiation).

for extraction of part of the radiation outside the cavity. It is either a grating operating by reflection or a beam separator (semi-transparent blade).

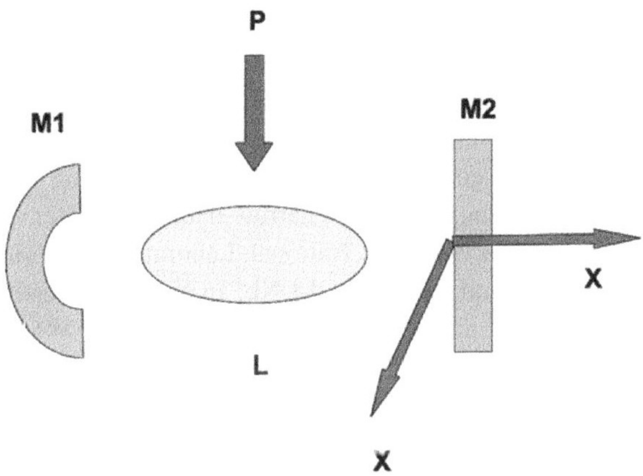

Figure 7.18. *Scheme of a cavity for an X-ray laser composed of two mirrors M1 and M2, containing the lasing medium L pumped by a laser with power P*

COMMENT ON FIGURE 7.18.– *The mirror M1 sends the X-ray radiation from medium L, potentially with focusing; the mirror M2 reflects the X-ray radiation and performs extraction of part of this radiation either by transmission (semi-transparent plate) or diffraction (reflection grating).*

These cavities based on multilayer Bragg structures have met expectations and enabled this type of laser to be developed. Thanks to a second pass through the lasing medium caused by the reflection at normal incidence of a Bragg multilayer mirror, an increase in intensity by 120% was reported at a wavelength of 18.2 nm (Ceglio et al. 1988). The main challenge faced was to create deposits capable of resisting the high power of radiation (>10^{10} W/cm^2) that was likely to irreversibly damage these optics. In fact, in a horizontal pumping scheme where the cavity mirror is placed a few centimeters from the plasma brought to a high temperature, the reflector is destroyed in a single laser shot, and in the subsequent X-ray laser generations, the amplification is completed in a single pass (Suckewer and Jaeglé 2009). One of the areas of development in this field is the creation of a table top laser that operates in the so-called "water window" domain, where potential applications are significant, primarily in the field of biology. Devices such as these, using methods that allow lasing at shorter wavelengths than the first lasers, for example flying mirrors (see Chapter 1), have been created (Esirkepov et al. 2009) and have benefitted from multilayer optical devices adapted to this spectral domain.

Among the main applications of X-ray lasers are microscopy and interferometry in the soft X-ray radiation domain and holography benefitting from progress in the field of artificial Bragg optics. By way of an example, we can cite reflection microscopes use Schwarzschild objectives equipped with Bragg multilayer mirrors operating under quasi normal incidence (Uspenskii et al. 1998). A holographic device, with Gabor inline[16] geometry modified by the inclusion of a Bragg mirror, composed of a multilayer stack used as a narrow bandwidth filter, was created by a group from the Lawrence Livermore National Laboratory (LLNL) in the United States (Trebes et al. 1987) and, also at the LLNL, an X-ray laser operating at 15.5 nm has enabled the density of a plasma to be probed using a Mach–Zehnder X-ray interferometer comprising several multilayer Bragg structures (Da Silva et al. 1995).

7.6.3. *Elements generating sources of X-ray radiation*

Different types of phenomena have been used or are envisaged to generate X-ray radiation using artificial Bragg structures. Those which have resulted in sources, which have been tested and characterized, are primarily based on quasi-Cherenkov phenomena using monoenergetic electron beams in interaction with a periodic structure playing the role of radiator. Generally, these sources emit quasi monochromatic radiation in a given direction in the form of an angularly narrow brush. To explain the quasi-Cherenkov characteristic, we can refer to Figure 7.19.

An electron traverses the periodic structure along a trajectory considered to be rectilinear, with a speed $\mathbf{v}$. Its environment is periodically modified based on a time period $\Delta t = \frac{2\pi}{|\mathbf{G.v}|}$ and spatial period $\Delta \mathbf{z} = \frac{2\pi}{|\mathbf{G.v}|}\,\mathbf{v}$. The time τ between emission of the radiation at the point $\mathbf{z}(t)$ and the point $\mathbf{z}(t) + \Delta \mathbf{z}$ is given by:

$$\tau = \Delta t - n\,\mathbf{k}.\Delta \mathbf{z} = (1 - n\,\mathbf{k}.\mathbf{v})\,\Delta t \qquad [7.9]$$

with $\mathbf{k}$ being the unit vector in the direction of observation of the radiation and n being the average refraction index. The phase shift ϕ between the two waves with frequency ω emitted respectively in $\mathbf{z}(t)$ and $\mathbf{z}(t) + \Delta \mathbf{z}$ is given by:

$$\phi = \omega\,\tau = \omega(1 - n\,\mathbf{k}.\mathbf{v})\,\Delta t \qquad [7.10]$$

16. This is geometry for holography where the source, the diffractive lens and the recording medium are aligned.

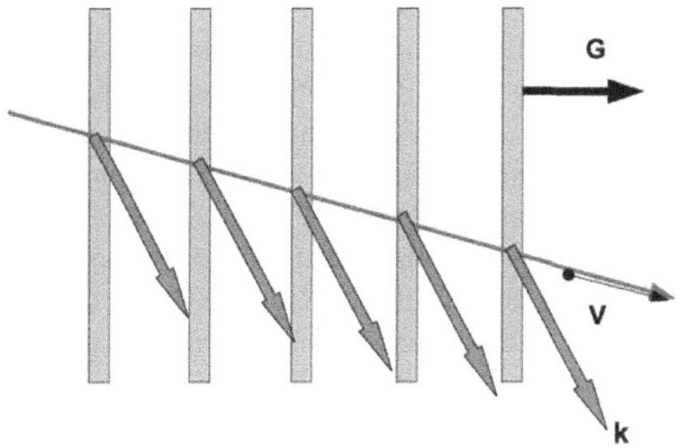

Figure 7.19. *Schematic diagram of the coherent electromagnetic emission generated by a charged particle, traversing at the speed **v** along the rectilinear trajectory represented by the red line a periodic structure represented by the blue bars*

COMMENT ON FIGURE 7.19.– *The vector **G** is the vector of the reciprocal network associated with the periodicity of the structure; the direction of emission is indicated by green arrows orientated according to the unit vector **k**.*

Significant emission is observed following a constructive interference phenomenon when $\phi = 2\,\pi\,p, p\,\in\,\mathbb{N}$, which leads to:

$$\omega(1 - n\,\mathbf{k}.\mathbf{v}) = p\,\mathbf{v}.\mathbf{G} \qquad [7.11]$$

which can be expressed by the fact that, in the direction given by the angle Θ counted with respect to the trajectory of the electron, resonant emission of radiation of frequency ω_p occurs, which is given by:

$$\omega_p = \frac{2\,\pi\,p}{|\Delta z|}\,\frac{|\mathbf{v}|}{1-n\,|\mathbf{v}|\,cos(\Theta)} \qquad [7.12]$$

When n is greater than 1 and $|\mathbf{v}| > 1/n$, we obtain the so-called Cherenkov emission corresponding to $p = 0$ and to the condition $n\,|\mathbf{v}|\,cos(\Theta) = 1$. In the case where the refraction index varies periodically in space, we obtain a quasi-Cherenkov emission phenomenon. Resonant transition radiation and parametric radiation, sometimes known as "quasi-Cherenkov", are associated with this type of phenomenon.

7.6.3.1. *Resonant transition radiation*

Basically, transition radiation occurs when an electrically charged particle, generally an electron, traverses an interface separating two optical media with different dielectric constants. The phenomenon was theorized by Ginzburg and Frank (1945) and Ginzburg (1996), and then the modeling was extended to the case of periodic media by M.L. Ter-Mikealyan (1961). The potential of this phenomenon to produce X-ray radiation under resonance conditions, in other words, by constructive interferences of waves emitted at the interfaces, was rapidly recognized, and experiments using radiators formed from material sheets arranged periodically have confirmed the usefulness of this approach in creating a source of X-ray radiation named X-ray resonant transition radiation (X-RTR) (Cherry 1978; Piestrup et al. 1992; Yajima et al. 1999). The difficulty of creating the correct periodicity with such radiators has resulted in different teams proposing and modeling the use of periodic multilayers at nanometric scale, such as the radiator to obtain X-RTR (Law and Kaplan 1987; Pardo and André 1989; Dubovikov 1994; Kaplan et al. 1995; André et al. 1999; Pardo and André 2002; Nasonov et al. 2003).

The use of a horizontal period gradient multilayer was proposed in order to create a source with tunable energy (André et al. 2003). An experiment performed by K. Yamada et al. (1999) using a periodic multilayer composed of Ni/C bilayers, subject to a 15 MeV electron beam, gave resonant emission of soft X-ray radiation in agreement with theoretical models. However, this experiment does not involve Bragg diffraction. This being said, certain models mention diffracted transition radiation (DTR) and discuss the distinction between this type of phenomenon and parametric radiation (Caticha 1989; Blazhevich et al. 2012).

Although Bragg diffraction is not involved in X-RTR, it seems important to mention that several authors have envisaged the possibility of creating an X-ray laser from the *stimulated* RTR by enclosing the radiator into a designated cavity (Piestrup and Finman 1983; Reid and Piestrup 1991), which could be made up of artificial Bragg structures according to a scheme close to the one represented in Figure 7.19. We return to this point in the perspectives in Chapter 8.

7.6.3.2. *Parametric radiation*

There is another quasi-Cherenkov radiation, generally denoted as parametric X-ray radiation (PXR) when it occurs in the X-ray radiation domain. This involves Bragg diffraction, in contrast to X-RTR. This radiation was initially observed using natural crystals (diamond, silicon, etc.) from the interaction of relativistic electrons at the 900 MeV synchrotron in Tomsk in the USSR (Baryshevsky et al. 1985; Adishchev et al. 1989, 1990). These experiments were followed by a series of other experiments using less energetic electrons. The phenomenon was modeled in the 1970s by Baryshevsky and Feranchuk (1972) and Garibyan and Yang (1972). Qualitatively, parametric radiation can be interpreted as a result of the Bragg diffraction of virtual photons[17] associated with the electron interacting with the periodic structure.

PXR, produced using crystals, generates hard X-ray radiation. The advent of artificial Bragg structures can be used to generate, under the same principle, softer X-ray radiation as proposed and modeled in various works (Pardo and André 2002). The teams in Tomsk were at the cutting edge of the discovery and characterization of the PXR (Kaplin et al. 2000), and it was demonstrated that this approach should allow soft X-ray radiation to be produced in the water window domain (Uglov et al. 2016), provided that the quality of the radiator is good enough, as shown in Figure 7.20.

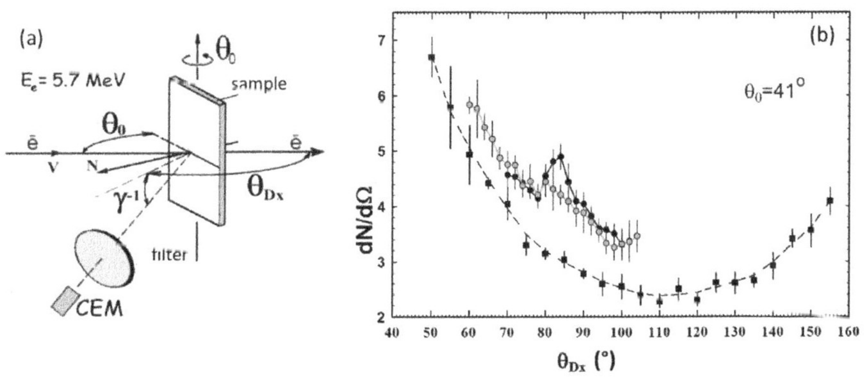

Figure 7.20. *PXR experiment. For a color version of this figure, see www.iste.co.uk/andre/xrayradiation.zip*

COMMENT ON FIGURE 7.20.– *(a) Scheme and geometry of the PXR experiment: a 5.7 MeV electron beam is incident on the target, a periodic structure of 100 Cr/Sc bilayers with a period of 2.34 nm is deposited on an Si_3N_4 membrane, and the radiation generated is detected by the CEM; (b) intensity detected as a function of the angle of detection: black dots: multilayer; blue points: membrane only; lower curve: background noise. According to Uglov et al. (2016)*[18].

17. Direct observation of virtual photons is not possible, but according to a model by von Weizsäcker and Williams (von Weizsäcker 1934), they are involved in a Coulomb interaction.
18. See doi: 10.1088/1742-6596/732/1/012017.

7.7. References

Adishchev, Y.N., Verzilov, V.A., Potylitsyn, A.P., Uglov, S.R., Vorobyev, S.A. (1989). Measurement of spectral and polarization characteristics of parametric X-rays in a Si crystal. *Nuclear Instruments and Methods in Physics Research Section B: Beam Interactions with Materials and Atoms*, 44(2), 130–136.

Adishchev, Y.N., Kaplin, V.V., Potylitsin, A.P., Uglov, S.R., Vorobiev, S.A., Babadzhanov, R.D., Verzilov, V.A. (1990). Influence of Kα absorption in (111) Ge crystal on spectral yield of parametric X-rays. *Physics Letters A*, 147(5), 326–328.

André, J.-M. (1998). Soft X-ray fluorescence spectrometer using a multilayer interferential transmission plate. *Review of Scientific Instruments*, 69(3), 1267–1269.

André, J.-M., Pardo, B., Bonnelle, C. (1999). XUV resonant transition radiations from periodic stratified media. *Physical Review E*, 60(1), 968–978.

André, J.-M., Benbalagh, R., Barchewitz, R., Ravet, M.-F., Raynal, A., Delmotte, F., Bridou, F., Julié, G., Bosseboeuf, A., Laval, R. et al. (2002). X-ray multilayer monochromator with enhanced performance. *Applied Optics*, 41(1), 239–244.

André, J.-M., Bonnelle, C., Pardo, B. (2003). Tunable and bright EUV sources using the interaction of medium–energy electron beams with periodic multilayer targets. *Nuclear Instruments and Methods in Physics Research Section A: Accelerators, Spectrometers, Detectors and Associated Equipment*, 498(1), 532–538.

André, J.-M., Jonnard, P., Michaelsen, C., Wiesmann, J., Bridou, F., Ravet, M.-F., Jérome, A., Delmotte, F., Filatova, E.O. (2005). La/B4C small period multilayer interferential mirror for the analysis of boron. *X-Ray Spectrometry*, 34(3), 203–206.

Aristov, V.V., Gaponov, S.V., Genkin, V.M., Gorbatov, Y.A., Martynov, V.V., Matveev, L.A., Salashchenko, N.N., Fraerman, A.A. (1986). Focusing properties of shaped multilayer X-ray mirrors. *JETP Letters*, 44(4), 265–267.

Baez, A.V. (1952). A study in diffraction microscopy with special reference to X-rays. *JOSA*, 42(10), 756–762.

Bajt, S., Chapman, H.N., Aquila, A., Gullikson, E. (2012). High-efficiency X-ray gratings with asymmetric-cut multilayers. *JOSA A*, 29(3), 216–230.

Bajt, S., Prasciolu, M., Fleckenstein, H., Domaracký, M., Chapman, H.N., Morgan, A.J., Yefanov, O., Messerschmidt, M., Du, Y., Murray, K.T. et al. (2018). X-ray focusing with efficient high-NA multilayer Laue lenses. *Light: Science & Applications*, 7(3), 17162.

Bakshi, V. (2018). *EUV Lithography*. SPIE Press, Bellingham [Online]. Available at: https://spie.org/Publications/Book/2305674.

Barchewitz, R. and André, J.-M. (1996). Structures multicouches gravées pour le rayonnement X-UV. *Le Journal de Physique IV*, 06(C4), C4-763–C4-772.

Baryshevsky, V.G. and Feranchuk, I.D. (1972). Transition radiation of γ rays in a crystal. *Journal of Experimental and Theoretical Physics*, 34(3), 502–504.

Baryshevsky, V.G., Danilov, V.A., Ermakovich, O.L., Feranchuk, I.D., Ivashin, A.V., Kozus, V.I., Vinogradov, S.G. (1985). Angular distribution of parametric X-rays. *Physics Letters A*, 110(9), 477–479.

Benbalagh, R., André, J.-M., Barchewitz, R., Jonnard, P., Julié, G., Mollard, L., Rolland, G., Rémond, C., Troussel, P., Marmoret, R. et al. (2005). Lamellar multilayer amplitude grating as soft-X-ray Bragg monochromator. *Nuclear Instruments and Methods in Physics Research Section A: Accelerators, Spectrometers, Detectors and Associated Equipment*, 541(3), 590–597.

Blazhevich, S.V., Kolosova, I.V., Noskov, A.V. (2012). Coherent X-ray radiation generated by a relativistic electron in an artificial periodic structure. *Journal of Experimental and Theoretical Physics*, 114(4), 547–554.

Borozdin, Y.É., Kazakov, E.D., Luchin, V.I., Salashchenko, N.N., Tolstikhina, I.Y., Chernov, V.V., Chkhalo, N.I., Shevel'ko, A.P., Yakushev, O.F. (2008). X-ray and vacuum-ultraviolet plasma spectroscopy with the use of new focusing multilayer structures. *JETP Letters*, 87(1), 27–29.

Brewer, C.A., Brizuela, F., Wachulak, P., Martz, D.H., Chao, W., Anderson, E.H., Attwood, D.T., Vinogradov, A.V., Artyukov, I.A., Ponomareko, A.G. et al. (2008). Single-shot extreme ultraviolet laser imaging of nanostructures with wavelength resolution. *Optics Letters*, 33(5), 518–520.

Caticha, A. (1989). Transition-diffracted radiation and the Cerenkov emission of X-rays. *Physical Review A*, 40(8), 4322–4329.

Cauchon, G., Pichet-Thomasset, M., Sauneuf, R., Dhez, P., Idir, M., Ollivier, M., Troussel, P., Boutin, J.-Y., Le Breton, J.-P. (1998). Imaging of laser produced plasma at 1.43 keV using Fresnel zone plate and Bragg–Fresnel lens. *Review of Scientific Instruments*, 69(9), 3186–3193.

Ceglio, N.M., Stearns, D.G., Hawryluk, A.M., Barbee, T.W., Danzmann, K., Kuhne, M., Mueller, P., Wende, B., Stearns, M.B., Petford-Long, A.K. et al. (1986). Soft X-ray laser cavities. *Le Journal de Physique Colloques*, 47(C6), C6-277–C6-286.

Ceglio, N.M., Stearns, D.G., Gaines, D.P., Hawryluk, A.M., Trebes, J.E. (1988). Multipass amplification of soft X-rays in a laser cavity. *Optics Letters*, 13(2), 108–110.

Celliers, P., Weber, F., Silva, L.B.D., Barbee, T.W., Cauble, R., Wan, A.S., Moreno, J.C. (1995). Fringe formation and coherence of a soft-X-ray laser beam illuminating a Mach–Zehnder interferometer. *Optics Letters*, 20(18), 1907–1909.

Cherry, M.L. (1978). Measurements of the spectrum and energy dependence of X-ray transition radiation. *Physical Review D*, 17(9), 2245–2260.

Cruddace, R.G., Barbee Jr., T.W, Rife, J.C., Hunter, W.R. (1990). Measurements of the normal-incidence X-ray reflectance of a molybdenum-silicon multilayer deposited on a 2000 l/mm grating. *Physica Scripta*, 41(4), 396.

Da Silva, L.B., Barbee, Jr., T.W., Cauble, R., Celliers, P., Ciarlo, D., Libby, S., London, R.A., Matthews, D., Mrowka, S., Moreno, J.C. et al. (1995). Electron density measurements of high density plasmas using soft X-ray laser interferometry. *Physical Review Letters*, 74(20), 3991–3994.

Delaboudinière, J.-P., Artzner, G.E., Brunaud, J., Gabriel, A.H., Hochedez, J.F., Millier, F., Song, X.Y., Au, B., Dere, K.P., Howard, R.A. et al. (1995). EIT: Extreme-ultraviolet imaging telescope for the SOHO mission. *Solar Physics*, 162(1), 291–312.

Delmotte, F., Ravet, M.-F., Bridou, F., Varnière, F., Zeitoun, P., Hubert, S., Vanbostal, L., Soullie, G. (2002). X-ray–ultraviolet beam splitters for the Michelson interferometer. *Applied Optics*, 41(28), 5905–5912.

DiCicco, D.S., Kim, D., Rosser, R., Suckewer, S. (1992). First stage in the development of a soft-X-ray reflection imaging microscope in the Schwarzschild configuration using a soft-X-ray laser at 18.2 nm. *Optics Letters*, 17(2), 157–159.

Döhring, T., Stollenwerk, M., Gong, Q., Proserpio, L., Winter, A., Friedrich, P. (2015). The challenge of developing thin mirror shells for future X-ray telescopes. *Proceedings of SPIE*, 9628, 17–24.

Dubovikov, M.S. (1994). Transition radiation and Bragg resonances. *Physical Review A*, 50(3), 2068–2074.

Erdmann, A. (2021). *Optical and EUV Lithography: A Modeling Perspective*. SPIE Press, Bellingham [Online]. Available at: https://www.spiedigitallibrary.org/ebooks/PM/Optical-and-EUV-Lithography-A-Modeling-Perspective/eISBN-9781510639027/10.1117/3.2576902 [Accessed 28 March 2023].

Erko, A. (1994). High resolution Bragg-Fresnel optics for nanometer scale methods with soft X-rays. *Le Journal de Physique IV*, 04(C9), C9-251.

Esirkepov, T.Z., Bulanov, S.V., Kando, M., Pirozhkov, A.S., Zhidkov, A.G. (2009). The flying mirror: Future brightest X-ray and gamma-ray source. *Proceedings of SPIE*, 7359, 75–85.

Fechtchenko, R.M., Vinogradov, A.V., Voronov, D.L. (2002). Optical properties of sliced multilayer gratings. *Optics Communications*, 210(3), 179–186.

Garibyan, G.M. and Yang, C. (1972). Quantum microscopic theory of radiation by a charged particle moving uniformly in a crystal. *JETP*, 34(3), 495–501.

Ginzburg, V.L. (1996). Radiation by uniformly moving sources (Vavilov–Cherenkov effect, transition radiation, and other phenomena). *Physics-Uspekhi*, 39(10), 973.

Ginzburg, V.L. and Frank, I.M. (1945). Radiation of a uniformly moving electron due to Its transition from one medium to another. *Journal Physics (USSR)*, 9, 353–362.

Giulietti, D. and Gizzi, L.A. (1998). X-ray emission from laser-produced plasmas. *La Rivista del Nuovo Cimento*, 21(10), 1–93.

Gluskin, E.S., Gaponov, S.V., Dhez, P., Ilyinsky, P.P., Salashchenko, N.N., Shatunov, Y.M., Trakhtenberg, E.M. (1986). A polarimeter for soft X-ray and VUV radiation. *Nuclear Instruments and Methods in Physics Research Section A: Accelerators, Spectrometers, Detectors and Associated Equipment*, 246(1), 394–396.

Haga, T., Tinone, M.C.K., Shimada, M., Ohkubo, T., Ozawa, A. (1998). Soft X-ray multilayer beam splitters. *Journal of Synchrotron Radiation*, 5(3), 690–692.

Hambach, D., Schneider, G., Gullikson, E.M. (2001). Efficient high-order diffraction of extreme-ultraviolet light and soft X-rays by nanostructured volume gratings. *Optics Letters*, 26(15), 1200–1202.

von Hámos, L. (1932). Röntgenspektroskopie und Abbildung mittels gekrümmter Kristallreflektoren. *Naturwissenschaften*, 20(38), 705–706.

Harada, T. (1990). Design and application of a varied-space plane grating monochromator for synchrotron radiation. *Nuclear Instruments and Methods in Physics Research Section A: Accelerators, Spectrometers, Detectors and Associated Equipment*, 291(1), 179–184.

Harrison, F.A., Craig, W.W., Christensen, F.E., Hailey, C.J., Zhang, W.W., Boggs, S.E., Stern, D., Cook, W.R., Forster, K., Giommi, P. et al. (2013). The Nuclear Spectroscopic Telescope Array (NuSTAR) high-energy X-ray mission. *The Astrophysical Journal*, 770(2), 103.

Hawryluk, A.M., Ceglio, N.M., Stearns, D.G., Danzmann, K., Kuhne, M., Muller, P., Wende, B. (1987). Soft X-ray beamsplitters and highly dispersive multilayer mirrors for use as soft X-ray laser cavity components. *Proceedings of SPIE*, 688, 81.

Hecquet, C., Delmotte, F., Ravet-Krill, M.-F., de Rossi, S., Jérome, A., Bridou, F., Varnière, F., Meltchakov, E., Auchère, F., Giglia, A. et al. (2009). Design and performance of two-channel EUV multilayer mirrors with enhanced spectral selectivity. *Applied Physics A*, 95(2), 401–408.

Heilmann, R.K., Bruccoleri, A.R., Burwitz, V., DeRoo, C., Garner, A., Günther, H.M., Gullikson, E.M., Hartner, G., Hertz, E., Langmeier, A. et al. (2022). X-ray performance of critical-angle transmission grating prototypes for the Arcus mission. *The Astrophysical Journal*, 934(2), 171.

Hombourger, C., Jonnard, P., André, J.-M., Chauvineau, J.-P. (1999). Use of layered synthetic microstructures for the quantitative X-ray analysis of light elements. *X-Ray Spectrometry*, 28(3), 163–167.

Idir, M., Mirone, A., Cauchon, G., Dhez, P. (1997). Optiques de Bragg-Fresnel. Principes et applications. *Annales de Physique*, 22(C1), 146.

Jakobsen, A.C., Ferreira, D.D.M., Christensen, F.E., Shortt, B., Collon, M., Ackermann, M.D. (2011). Preliminary coating design and coating developments for ATHENA. *Proceedings of SPIE*, 8147, 290–297.

Jark, W. (1986). Enhancement of diffraction grating efficiencies in the soft X-ray region by a multilayer coating. *Optics Communications*, 60(4), 201–205.

Joensen, K.D., Voutov, P., Szentgyorgyi, A., Roll, J., Gorenstein, P., Hoghoj, P., Christensen, F.E. (1995). Design of grazing-incidence multilayer supermirrors for hard-X-ray reflectors. *Applied Optics*, 34(34), 7935–7944.

Jonnard, P., Le Guen, K., André, J.-M. (2009). High-resolution X-ray analysis with multilayer gratings. *X-Ray Spectrometry*, 38(2), 117–120.

Jonnard, P., Le Guen, K., André, J.-M., Coudevylle, J.-R., Isac, N. (2012). An etched multilayer as a dispersive element in a curved-crystal spectrometer: Implementation and performance. *X-Ray Spectrometry*, 41(5), 308–312.

Jonnard, P., Wu, M., André, J.-M., Le Guen, K., Wang, Z., Huang, Q., Vickridge, I., Schmaus, D., Briand, E., Steydli, S. et al. (2018). Note: Observation of the angular distribution of an X-ray characteristic emission through a periodic multilayer. *Review of Scientific Instruments*, 89(9), 096109.

Kang, H.C., Yan, H., Chu, Y.S., Lee, S.Y., Kim, J., Nazaretski, E., Kim, C., Seo, O., Noh, D.Y., Macrander, A.T. et al. (2013). Oxidation of PtNi nanoparticles studied by a scanning X-ray fluorescence microscope with multi-layer Laue lenses. *Nanoscale*, 5(16), 7184–7187.

Kaplan, A.E., Law, C.T., Shkolnikov, P.L. (1995). X-ray narrow-line transition radiation source based on low-energy electron beams traversing a multilayer nanostructure. *Physical Review E*, 52(6), 6795–6808.

Kaplin, V.V., Uglov, S.R., Zabaev, V.N., Piestrup, M.A., Gary, C.K., Nasonov, N.N., Fuller, M.K. (2000). Observation of bright monochromatic X-rays generated by relativistic electrons passing through a multilayer mirror. *Applied Physics Letters*, 76(24), 3647–3649.

Keski-Kuha, R.A.M. (1984). Layered synthetic microstructure technology considerations for the extreme ultraviolet. *Applied Optics*, 23(20), 3534–3537.

Khan Malek, C., Madouri, A., Rivoira, R., Susini, J., Ouahabi, M., Ladan, F.R., Lepêtre, Y., Barchewitz, R. (1989). Microfabrication technologies applied to the production of large area stress relieved multilayer beamsplitters. *Proceedings of SPIE*, 1160, 49.

Kirz, J. (1974). Phase zone plates for X-rays and the extreme UV. *JOSA*, 64(3), 301–309.

Kortright, J.B. and Underwood, J.H. (1990). Multilayer optical elements for generation and analysis of circularly polarized X-rays. *Nuclear Instruments and Methods in Physics Research Section A: Accelerators, Spectrometers, Detectors and Associated Equipment*, 291(1), 272–277.

Kortright, J.B., Kimura, H., Nikitin, V., Mayama, K., Yamamoto, M., Yanagihara, M. (1992). Soft X-ray (97 eV) phase retardation using transmission multilayers. *Applied Physics Letters*, 60(24), 2963–2965.

Kortright, J.B., Rice, M., Carr, R. (1995). Soft-X-ray Faraday rotation at Fe L2,3 edges. *Physical Review B*, 51(15), 10240–10243.

Koyama, T., Takenaka, H., Ichimaru, S., Ohchi, T., Tsuji, T., Takano, H., Kagoshima, Y. (2011a). Development of multilayer Laue lenses; (1) Linear type. *AIP Conference Proceedings*, 1365(1), 24–27.

Koyama, T., Tsuji, T., Takano, H., Kagoshima, Y., Ichimaru, S., Ohchi, T., Takenaka, H. (2011b). Development of multilayer Laue lenses; (2) Circular type. *AIP Conference Proceedings*, 1365(1), 100–103.

Koyama, T., Takano, H., Konishi, S., Tsuji, T., Takenaka, H., Ichimaru, S., Ohchi, T., Kagoshima, Y. (2012). Circular multilayer zone plate for high-energy X-ray nano-imaging. *Review of Scientific Instruments*, 83(1), 013705.

Kozhevnikov, I.V., van der Meer, R., Bastiaens, H.M.J., Boller, K.-J., Bijkerk, F. (2010). High-resolution, high-reflectivity operation of lamellar multilayer amplitude gratings: Identification of the single-order regime. *Optics Express*, 18(15), 16234–16242.

Krastev, K., Le Guern, F., Coat, K., Barchewitz, R., André, J.-M., Ravet, M.-F., Cambril, E., Rousseaux, F., Davi, P. (1996). Spectral sampling of a soft X-ray tube emission by a lamellar multilayer grating in standard and conical mountings. *Nuclear Instruments and Methods in Physics Research Section A: Accelerators, Spectrometers, Detectors and Associated Equipment*, 368(2), 533–542.

Lang, J.C. and Srajer, G. (1995). Bragg transmission phase plates for the production of circularly polarized X-rays. *Review of Scientific Instruments*, 66(2), 1540–1542.

Law, C.T. and Kaplan, A.E. (1987). X-ray transition radiation in a solid-state superlattice: Photoabsorption, electron scattering, and radiation optimization. *Optics Letters*, 12(11), 900–902.

Le Guen, K., Benbalagh, R., André, J.-M., Coudevylle, J.-R., Jonnard, P. (2017). Design, development and applications of etched multilayers for soft X-ray spectroscopy. *The European Physical Journal Applied Physics*, 78(2), 20702.

Libby, S.B., Da Silva, L.B., Barbee, T.W., Cauble, R., Celliers, P., London, R.A., Matthews, D.L., Mrowka, S., Moreno, J.C., Ress, D. et al. (1996). The X-ray laser as a tool for imaging plasmas. *AIP Conference Proceedings*, 369(1), 737–742.

Lubov, M. and Goray, L. (2019). High-efficiency X-ray multilayer-coated blazed gratings with shifted boundaries. *Journal of Synchrotron Radiation*, 26(5), 1539–1545.

MacDonald, M.A., Schaefers, F., Pohl, R., Poole, I.B., Gaupp, A., Quinn, F.M. (2008). A W:B4C multilayer phase retarder for broadband polarization analysis of soft X-ray radiation. *Review of Scientific Instruments*, 79(2), 025108.

Martínez-Galarce, D.S., Soufli, R., Windt, D.L., Bruner, M.E., Gullikson, E.M., Khatri, S., Spiller, E.A., Robinson, J.C., Baker, S.L., Prast, E. (2013). Multisegmented, multilayer-coated mirrors for the Solar Ultraviolet Imager. *Optical Engineering*, 52(9), 095102.

Matthews, D.L., Hagelstein, P.L., Rosen, M.D., Eckart, M.J., Ceglio, N.M., Hazi, A.U., Medecki, H., MacGowan, B.J., Trebes, J.E., Whitten, B.L. et al. (1985). Demonstration of a soft X-ray amplifier. *Physical Review Letters*, 54(2), 110–113.

Meltchakov, E., Hecquet, C., Roulliay, M., De Rossi, S., Menesguen, Y., Jérome, A., Bridou, F., Varniere, F., Ravet-Krill, M.-F., Delmotte, F. (2009). Development of Al-based multilayer optics for EUV. *Applied Physics A*, 98(1), 111.

Meltchakov, E., Ziani, A., Auchere, F., Zhang, X., Roulliay, M., Rossi, S.D., Bourassin-Bouchet, C., Jérome, A., Bridou, F., Varniere, F. et al. (2011). EUV reflectivity and stability of tri-component Al-based multilayers. *Proceedings of SPIE*, 8168, 293–301.

Mirone, A., Idir, M., Soullie, G., Dhez, P. (1994). X-ray imaging with off-axis Bragg Fresnel multilayer lenses. *Le Journal de Physique IV*, 04(C9), 276.

Misture, S.T. (2008). Application of single and dual multilayer optics for powder X-ray diffraction. *X-Ray Optics and Instrumentation*, e408702.

Montel, M. (1957). The X-ray microscope with catamegonic roof-shaped objective. In *X-Ray Microscopy and Microradiography*, Cosslett, V.E., Engstrom, A., Pattee, H.H. (eds). Elsevier, Amsterdam, 117–185.

Murray, K.T., Pedersen, A.F., Mohacsi, I., Detlefs, C., Morgan, A.J., Prasciolu, M., Yildirim, C., Simons, H., Jakobsen, A.C., Chapman, H.N. et al. (2019). Multilayer Laue lenses at high X-ray energies: Performance and applications. *Optics Express*, 27(5), 7120–7138.

Nasonov, N.N., Kaplin, V.V., Uglov, S.R., Piestrup, M.A., Gary, C.K. (2003). X-rays from relativistic electrons in a multilayer structure. *Physical Review E*, 68(3), 036504.

Pardo, B. and André, J.-M. (1989). Transition radiation from periodic stratified structures. *Physical Review A*, 40(4), 1918–1925.

Pardo, B. and André, J.-M. (2002). Parametric and resonant transition radiation in periodic stratified structures. *Physical Review E*, 65(3), 036501.

Piestrup, M.A. and Finman, P. (1983). The prospects of an X-ray free electron laser using stimulated resonance transition radiation. *IEEE Journal of Quantum Electronics*, 19(3), 357–364.

Piestrup, M.A., Boyers, D.G., Pincus, C.I., Li, Q., Hallewell, G.D., Moran, M.J., Skopik, D.M., Silzer, R.M., Maruyama, X.K., Snyder, D.D. et al. (1992). Observation of soft-X-ray spatial coherence from resonance transition radiation. *Physical Review A*, 45(2), 1183–1196.

Polkonikov, V., Chkhalo, N., Pleshkov, R., Giglia, A., Rividi, N., Brackx, E., Le Guen, K., Ismail, I., Jonnard, P. (2021). Periodic multilayer for X-ray spectroscopy in the Li K range. *Applied Sciences*, 11(14), 6385.

Prasciolu, M., Haase, A., Scholze, F., Chapman, H.N., Bajt, S. (2015). Extended asymmetric-cut multilayer X-ray gratings. *Optics Express*, 23(12), 15195–15204.

Reid, M.B. and Piestrup, M.A. (1991). Resonance transition radiation X-ray lasers. *IEEE Journal of Quantum Electronics*, 27(11), 2440–2455.

Rife, J.C., Hunter, W.R., Barbee Jr., T.W., Cruddace, R.G. (1989). Multilayer-coated blazed grating performance in the soft X-ray region. *Applied Optics*, 28(15), 2984.

Rochus, P., Auchère, F., Berghmans, D., Harra, L., Schmutz, W., Schühle, U., Addison, P., Appourchuax, T., Aznar Cuadrado, R., Baker, D. et al. (2020). The Solar Orbiter EUI instrument: The extreme ultraviolet imager. *Astronomy & Astrophysics*, 642, A8.

Sammar, A. and André, J.-M. (1993). Diffraction of multilayer gratings and zone plates in the X-ray region using the Born approximation. *JOSA A*, 10(4), 600–613.

Schäfers, F., Mertins, H.-C., Gaupp, A., Gudat, W., Mertin, M., Packe, I., Schmolla, F., Fonzo, S.D., Soullié, G., Jark, W. et al. (1999). Soft-X-ray polarimeter with multilayer optics: Complete analysis of the polarization state of light. *Appl. Opt.* 38, 4074–4088. doi: 10.1364/AO.38.004074.

Schuster, M. and Gobel, H. (1995). Parallel-beam coupling into channel-cut monochromators using curved graded multilayers. *Journal of Physics D: Applied Physics*, 28(4A), A270.

Senf, F., Bijkerk, F., Eggenstein, F., Gwalt, G., Huang, Q., Kruijs, R., Kutz, O., Lemke, S., Louis, E., Mertin, M. et al. (2016). Highly efficient blazed grating with multilayer coating for tender X-ray energies. *Optics Express*, 24(12), 13220–13230.

Shimizu, K. and Omote, K. (2008). Multilayer optics for X-ray analysis. *Rigaku Journal*, 24(1), 1–9.

Smith, R.F., Dunn, J., Hunter, J.R., Nilsen, J., Hubert, S., Jacquemot, S., Remond, C., Marmoret, R., Fajardo, M., Zeitoun, P. et al. (2003). Longitudinal coherence measurements of a transient collisional X-ray laser. *Optics Letters*, 28(22), 2261–2263.

Sobierajski, R., Loch, R.A., van de Kruijs, R.W.E., Louis, E., von Blanckenhagen, G., Gullikson, E.M., Siewert, F., Wawro, A., Bijkerk, F. (2013). Mo/Si multilayer-coated amplitude-division beam splitters for XUV radiation sources. *Journal of Synchrotron Radiation*, 20(2), 249–257.

Sokolov, A., Huang, Q., Senf, F., Feng, J., Lemke, S., Alimov, S., Knedel, J., Zeschke, T., Kutz, O., Seliger, T. et al. (2019). Optimized highly efficient multilayer-coated blazed gratings for the tender X-ray region. *Optics Express*, 27(12), 16833–16846.

Suckewer, S. and Jaeglé, P. (2009). X-ray laser: Past, present, and future. *Laser Physics Letters*, 6(6), 411–436.

Suckewer, S., Skinner, C.H., Milchberg, H., Keane, C., Voorhees, D. (1985). Amplification of stimulated soft X-ray emission in a confined plasma column. *Physical Review Letters*, 55(17), 1753–1756.

Susini, J., Barchewitz, R., Marmoret, R., Rivoira, R., Lepêtre, Y. (1988). Semitransparent multilayer mirrors in the soft X-ray range. *Proceedings of SPIE*, 984, 263.

Ter-Mikaelyan, M.L. (1961). Emission of fast particles in a heterogeneous medium. *Nuclear Physics*, 24(1), 43–61.

Trebes, J.E., Brown, S.B., Campbell, E.M., Matthews, D.L., Nilson, D.G., Stone, G.F., Whelan, D.A. (1987). Demonstration of X-ray holography with an X-ray laser. *Science*, 238(4826), 517–519.

Uglov, S.R., Kaplin, V.V., Kubankin, A.S., André, J.-M., Le Guen, K., Jonnard, P., De Rossi, S., Meltchakov, E., Delmotte, F. (2016). Cr/Sc multilayer radiator for parametric EUV radiation in "water-window" spectral range. *Journal of Physics: Conference Series*, 732(1), 012017.

Uspenskii, Y.A., Levashov, V.E., Vinogradov, A.V., Fedorenko, A.I., Kondratenko, V.V., Pershin, Y.P., Zubarev, E.N., Fedotov, V.Y. (1998). High-reflectivity multilayer mirrors for a vacuum-ultraviolet interval of 35–50nm. *Optics Letters*, 23(10), 771–773.

Uzoma, P.C., Shabbir, S., Hu, H., Okonkwo, P.C., Penkov, O.V. (2021). Multilayer reflective coatings for BEUV lithography: A review. *Nanomaterials*, 11(11), 2782.

Voronov, D.L., Cambie, R., Feshchenko, R.M., Gullikson, E.M., Padmore, H.A., Vinogradov, A.V., Yashchuk, V.V. (2007). Development of an ultrahigh-resolution diffraction grating for soft X-rays. *Proceedings of SPIE*, 6705, 102–113.

Voronov, D.L., Warwick, T., Padmore, H.A. (2014). Multilayer-coated blazed grating with variable line spacing and a variable blaze angle. *Optics Letters*, 39(21), 6134–6137.

Wang, Y., Narayanan, S., Liu, J., Shu, D., Mashayekhi, A., Qian, J., Wang, J. (2007). A sagittally focusing double-multilayer monochromator for ultrafast X-ray imaging applications. *Journal of Synchrotron Radiation*, 14(1), 138–143.

Wang, Z., Shen, Z., Mu, B., Wang, X., Yang, X., Jiang, L., Qi, R., Wen, M., Zhang, Z., Ma, B. (2014). Development of the X-ray timing and polarization telescope optics. *Proceedings of SPIE*, 9144, 411–416.

von Weizsäcker, C.F. (1934). Ausstrahlung bei Stößen sehr schneller Elektronen. *Zeitschrift für Physik*, 88(9), 612–625.

Windt, D.L., Christensen, F.E., Craig, W.W., Hailey, C., Harrison, F.A., Jimenez-Garate, M., Kalyanaraman, R., Mao, P.H. (2000). Growth, structure, and performance of depth-graded W/Si multilayers for hard X-ray optics. *Journal of Applied Physics*, 88(1), 460–470.

Wurm, S. and Kemp, K. (2006). SEMATECH pushes extreme UV lithography forward. SPIE Newsroom [Online]. Available at: https://www.spie.org/news/0079-sematech-pushes-extreme-uv-lithography-forward [Accessed 28 March 2023].

Yajima, K., Awata, T., Koizumi, R., Imai, M., Itoh, A., Imanishi, N. (1999). Generation of quasimonochromatic transition radiation X-rays using a K-shell photoabsorption edge. *Nuclear Instruments and Methods in Physics Research Section A: Accelerators, Spectrometers, Detectors and Associated Equipment*, 435(3), 490–494.

Yamada, K., Hosokawa, T., Takenaka, H. (1999). Observation of soft X-rays of single-mode resonant transition radiation from a multilayer target with a submicrometer period. *Physical Review A*, 59(5), 3673–3680.

Yan, H., Maser, J., Macrander, A., Shen, Q., Vogt, S., Stephenson, G.B., Kang, H.C. (2007). Takagi-Taupin description of X-ray dynamical diffraction from diffractive optics with large numerical aperture. *Physical Review B*, 76(11), 115438.

Yang, S., Chen, S., Lin, C. (2021). Design of reflective phase retarders in the extreme ultraviolet based on chirped Mo/Si multilayer mirrors. *Journal of Synchrotron Radiation*, 28(5), 1437–1443.

8

Perspectives and Conclusion

Jean-Michel ANDRÉ and Philippe JONNARD
LCPMR, CNRS, Sorbonne Université, Paris, France

8.1. Perspectives for fabrication and characterization

8.1.1. *Fabrication*

Using techniques for the deposition of films of nanometric thickness, combined with various methods of optimization (ionic polishing, nitriding, etc.), we have almost reached the limits of the possibilities offered by the current technology. It is not very likely that the performances in terms of reduction of interfacial roughness or interdiffusion phenomena can be considerably improved. Certainly, each new type of deposit requires a series of specific studies. This is the case, for example, for optics from the new generation of lithography with short wavelengths, which implies a long and onerous development phase. Nevertheless, progress must be made to increase the temporal stability of deposits for industrial applications (lithography, for example) and their resistance to damage, in particular for their use under the flow of intense radiation delivered by the new sources (X-ray laser, XFEL, etc.).

However, the most significant progress should relate to the technologies that allow multilayer deposits to be created without deteriorating their initial performances. Bragg–Fresnel lenses and etched diffraction gratings in a multilayer structure require fabrication processes in several stages, which tend to degrade the reflecting properties of a periodic stack. In addition, these microfabrication techniques remain difficult and expensive. Progress must therefore be made to improve the performances of these combined structures and to simplify their fabrication and reduce their cost.

In this context, we insist on the development of volume diffraction gratings operating at Bragg resonance (see Chapter 7). There could be many applications of these in spectroscopy, for example for analysis of light elements such as lithium, and they could be inserted relatively easily into instruments such as Castaing electron microprobes or electron microscopes. Nevertheless, these gratings require very high aspect ratios that are difficult to achieve. Here again, significant progress is necessary and will require participation from specialists in microfabrication. Another domain that will require a contribution from specialists, in this case opticians, is the creation of instruments that require curved substrates with higher tolerance criteria in terms of precision, surfacing, etc. Future space telescopes such as the SOLAR-C mission (see hereafter) fall into this category of instruments; a characterization of the surfaces of waves at the operating wavelengths will rapidly become necessary to achieve the required performances.

8.1.2. *Characterization*

Measurement of the performances of sources and X-ray optics is undertaken primarily in terms of intensity (illumination, reflectance, transmittance, etc.), while metrology, concerning in particular the in situ phase and wavefront, is still used very little, although some works on this subject have been produced (Le Pape et al. 2001; Idir et al. 2022). Wavefront analysis methods using Hartmann devices have also been developed on XFEL or synchrotron lines (Mercère et al. 2004; Bachelard et al. 2011). However, measurement of the wavefront is critical for the development of precision instruments, especially to reduce aberrations by acting on the fabrication processes or by developing active correction techniques (Nematollahi et al. 2019; Sutter et al. 2019). The specialized literature does not report developed metrology of the wavefront in relation to optics based on artificial Bragg structures. Recently, a laboratory device was fabricated which can routinely perform wavefront analyses using a ptychographic[1] technique and used to characterize the aberrations of a multilayer Laue lens (see Chapter 6) (Dresselhaus et al. 2022). We note that Bragg

1. Ptychography is a lens-free imaging technique that uses different diffraction patterns obtained by moving the object with high coverage. From the intensity and the phase deduced from the interferences between the various diffraction patterns, the real image is deduced generally using highly elaborate mathematical methods (Chapman 1996). This method has been adapted to the X-ray radiation domain and associated with tomography.

ptychography is being developed to analyze crystals (Hruszkewycz et al. 2017), and we could envisage its extension to artificial Bragg structures even though this method still requires a large amount of time for measurement and calculation. The panchromatic Talbot effect[2] was used by researchers at ONERA in collaboration with the SOLEIL synchrotron in France to develop phase imaging in the X-ray range and could be implemented to characterize artificial Bragg mirrors. It is probable that this type of metrology will be developed in the future with the increasing demand for high-precision X-ray optical instruments.

8.2. Perspectives for applications

8.2.1. *Optics and imaging*

The availability of ultrashort X-ray sources, which could deliver pulses of the order of a femtosecond for XFEL or an attosecond for high-harmonic generation (HHG) sources, means that dynamic analysis or control of quantum processes such as chemical reactions can be envisaged. These operations can require creation of light pulses using a device known as an optical pulse shaper (OPS). Among these, the most widely used in the long wavelength domain is the "4f shaper" (Haner and Warren 1988; Weiner et al. 1988) schematized in Figure 8.1(a), which includes two diffraction gratings (G1, G2), two lenses (L1, L2) and an amplitude and phase modulator (M). This device has been adapted in the EUV domain using optics based on artificial Bragg structures, as shown in Figure 8.1(b) (Aquila et al. 2011). Spectral scattering is performed using a truncated MG grating in a periodic multilayer (see Chapter 7). A Bragg reflecting structure M equipped with a mask provides the amplitude and phase modulation, and a cylindrical mirror with a multilayer coating operating under wide-band Bragg reflection collimates or focuses the beam.

It is highly probable that this type of device will be developed under various configurations and will contribute to pushing the frontier of analysis and control of the quantum processes using ultrashort pulses towards high-energy domains.

2. In 1836, W.H.F. Talbot discovered the eponymous effect that consists of spatial repetitions of the image of a diffractive element (grating) under monochromatic illumination. This phenomenon occurs periodically at discrete distances (Talbot plans), which depend on the wavelength and the grating period. In the case of polychromatic illumination, the interference figure becomes invariant by propagation (Guérineau et al. 2000). This panchromatic Talbot effect was demonstrated experimentally in the X-ray domain by Rizzi et al. (2011) and allows phase imaging to be performed in this spectral domain.

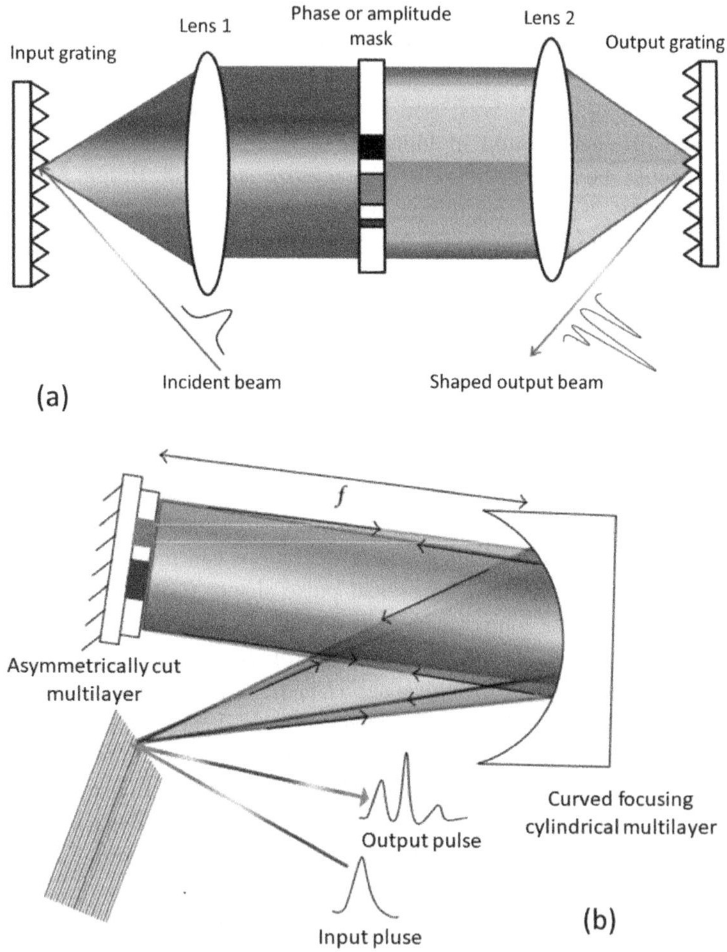

Figure 8.1. *(a) Schematic diagram of a "4f shaper" as created in the visible domain, according to Nag et al. (2010)*[3]*; (b) transposition to the EUV domain using artificial Bragg structures, according to Aquila et al. (2011)*[4]*. For a color version of this figure, see www.iste.co.uk/andre/xrayradiation.zip*

A domain in which there will most probably be a significant development, with a considerable associated budget, is short wavelength lithography. In Chapter 7, EUV lithography (EUVL) was presented, but in order to continue to reduce the

3. See doi: 10.1063/1.3276682.
4. See doi: 10.1155/2011/417075.

dimensions of the electronic circuits, it is necessary to use higher energy radiation than that available in EUV. In the context of the project *Beyond Extreme Ultraviolet Lithography*, research has focused on the use of radiation with a wavelength close to 6.5–7 nm with which patterns of the order of 10 nm can be made. This spectral domain corresponds to the anomalous region of boron; numerous investigations have been completed on multilayer mirrors made from this element. Reflectances above 80% are theoretically possible with mirrors made of La/B bilayers. Different techniques (nitriding, etc.) are being studied to ensure stability of these multilayers and reduce the effects of defects such as interface roughness or interdiffusion (Uzoma et al. 2021).

Another domain with extensive prospects involving international collaboration with large budgets is astrophysics and in particular the domain concerning the Sun. Our planet Earth, in the same way as the other planets in orbit around the Sun, is affected by solar winds, radiation and particles from the Sun, with sometimes dramatic consequences such as electrical system breakdowns. Observation of the Sun, and understanding of the physics that governs it, is therefore an important subject. As seen in Chapter 7, many solar exploration space missions using various observation instruments (telescopes, spectrographs, etc.) have been set up since the end of the 20th century. Imaging and spectroscopy in the EUV domain are high performance tools in this context because the solar corona and the chromosphere are made up of a plasma brought to high temperature that emits radiation in this spectral domain.

As an example of future space missions with instruments operating in the EUV domain and using artificial Bragg structures, we can cite the SOLAR-C mission. This project, led by the Japanese space agency JAXA, is based on a collaboration involving American and European laboratories and companies under the guidance of the respective space agencies (NASA and ESA). On board the SOLAR-C spaceship, there is an instrument called the EUVST characterized by the use of a minimum number of optical elements, more specifically a primary mirror that collects and focuses EUV radiation towards a slit, and a diffraction grating, both equipped with artificial Bragg structures, as shown in Figure 8.2. Therefore, the dispersive structure (grating) would be made up of an ellipsoid containing an etched Bragg multilayer structure to obtain a diffraction grating. SOLAR-C opens the door to a new generation of instruments for observation of the Sun, of a smaller size, but offering spatial, spectral and temporal resolutions that are a clear improvement on previous missions.

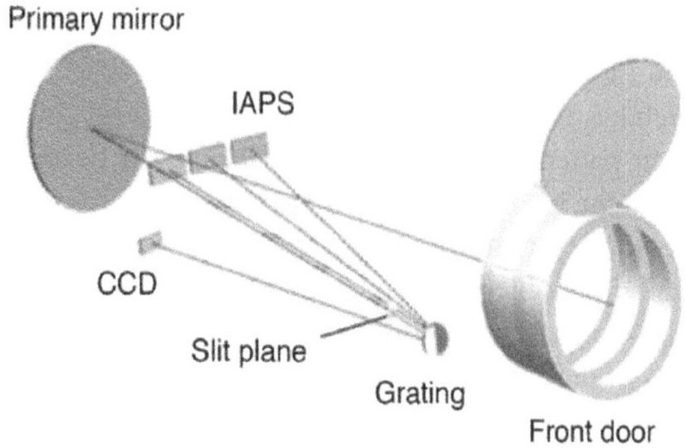

Figure 8.2. *Scheme for the EUVST instrument of the Solar-C mission with the primary parabolic mirror focusing the radiation in the plane of the slit; an etched concave grating in a Bragg structure deposited on an elliptical substrate disperses the radiation*[5]

An area that presents a certain futuristic aspect, but which has real potential, is communications in the domain of X-ray radiation (X communication, XCOM). This concept was put forward by K. Gendreau at the United States space agency (NASA), in the context of the communication blackout that occurs when a spacecraft enters the Earth's atmosphere, employing radiofrequency bands currently used. The frequencies of X-rays of the order of 10^{17} to 10^{18} Hz do not encounter this problem, and carrier waves with photons of energy near 10 keV could be envisaged for the purpose of providing communications with the spacecraft and interstellar communications (Hang et al. 2019).

In addition, XCOM offers potentiality for high-speed communication in agreement with Shannon's theory. Among the technological problems that must be resolved, there is the creation of quite powerful modulated X sources and optics with the capacity to form the beam at emission and to control reception at detection. Devices based on artificial Bragg structures with in particular super-mirror antenna (see Chapter 4) have been envisaged to provide these optical functions (Wang et al. 2021). It is probable that if there is a real interest for XCOM projects among major space agencies, there will be a significant development in this area of optics.

A domain that is difficult to access in the short wavelength domain, namely nonlinear optics, could make use of Bragg mirror reflection under quasi-normal incidence for which energy is tunable. More precisely, optical effects resulting from the mix of four waves (phase conjugation, parametric oscillation, etc.) proposed and

5. According to: https://solar-c.nao.ac.jp/en/instruments.html.

analyzed by Yariv and Pepper (1977) and observed at long wavelengths (Pepper et al. 2008) could be obtained in the EUV domain using mirrors according to a scheme provided by these authors (see Figure 8.3).

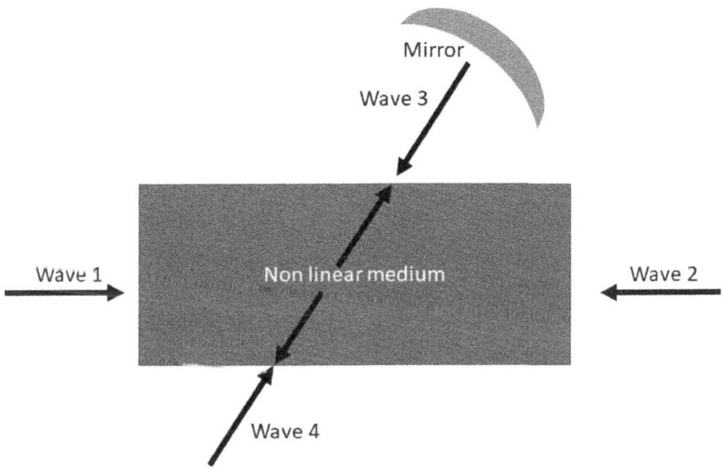

Figure 8.3. *Scheme of a system with a mix of four waves using an external mirror, which reduces the oscillation threshold with respect to a system without a mirror*

8.2.2. *Sources*

Artificial Bragg structures present various prospects in the field of X-ray sources, in particular regarding lasers in this spectral domain. They can be envisaged as an optical element to create external or internal cavities. The idea, proposed by A.L. Schawlow and C.H. Townes in 1954 (Schawlow and Townes 1958), of inserting the lasing medium into a resonant cavity is the basis for the majority of lasers created currently. The first free electron lasers operating at long wavelengths (infrared, visible, ultraviolet) also used optical cavities (Madey 1971; Deacon et al. 1977; Litvinenko et al. 1995). These cavities play an essential role in the properties of lasers, in particular in ensuring temporal (longitudinal) coherence. However, even though the concept of an oscillator was considered, during their design at the start of the 1990s, as the most direct method for achieving lasing at short wavelengths (X-ray radiation of EUV with hard X-rays) (Liss et al. 2000; Shvyd'ko et al. 2003), cavities of this kind in this spectral domain were then considered impossible to create. Consequently, the XFELs currently in service do not use these cavities and therefore suffer from several disadvantages:

– their size (and therefore their cost) related to the fact that they are supplied by low spontaneous emission, which requires long sections of undulator to achieve saturation;

– their low degree of monochromaticity of the order of 1%;

– the lack of longitudinal coherence given that a pulse includes hundreds or tens of thousands of longitudinal modes;

– the presence of fluctuations in time and intensity.

The situation is evolving thanks to methods that involve optical devices. Currently, the most developed method is so-called self-seeding, which was mentioned in Chapter 1. It consists of inserting an appropriate optical device after the first undulator(s) operating at a low gain regime where the SASE phenomenon occurs. This optical device "cleans" the radiation by filtering, conferring better characteristics to it in terms of temporal coherence, monochromaticity, etc. Amplification of this filtered radiation is then provided up to saturation by a section of undulator that operates as a radiator. The optical devices have until now been monochromators, natural crystals in the hard X-ray domain or diffraction gratings in the soft radiation domain.

Artificial Bragg structures could be used in this situation: it has been proposed that they should be used for this purpose, in backscattering mode in the soft radiation domain (André et al. 2023), which would mean that the use of gratings with a variable step, which are costly and difficult to implement, could be avoided. The problem of damage of these optics remains an obstacle that must be overcome. It is still true that with self-seeding, the pulses have not reached the transform-limited regime and continue to suffer from fluctuations in intensity. It therefore appears useful to return to the initial principle of the optical cavity. To date, several proposals have been made based on this principle as part of an approach known as the cavity-based X-ray FEL (CBXFEL). We can cite the X-ray regenerative amplifier FEL (XRAFEL) project by Z. Huang and R.D. Ruth in 2006 (Huang and Ruth 2006) and the XFELO project by K.-J. Kim et al. (Kim et al. 2008; Kim and Shvyd'ko 2009). These two types of project make use of Bragg reflection by means of monochromatic crystals. The diamond is the crystal that is currently prioritized (Shvyd'ko et al. 2010) due to its thermal stability and diffractive properties. Various XFEL centers (LCLS, European XFEL) are currently at the stage of development or proof of principle of devices based on the CBXFEL approach (Rauer et al. 2023). Figure 8.4 shows the principle of these CBXFELs in so-called zig-zag geometry, offering more degrees of freedom than backscattering in terms of tunability.

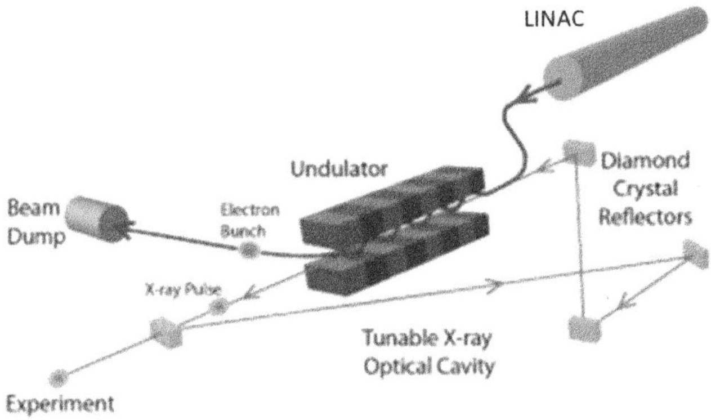

Figure 8.4. *Scheme of an oscillator for XFELs in the hard X-ray radiation domain in so-called zig-zag geometry (Kim et al. 2008); in the soft X-ray and FUV radiation domain, the reflectors could be replaced by artificial Bragg mirrors with high reflecting power and high stability. According to Krzywiński et al. (2019)*[6]

X-ray lasers based on population inversion are also involved in this trend (Halavanau et al. 2020) and current technologies appear to allow schemes announced in the 1960s and 1970s to come to fruition (Duguay and Rentzepis 1967; Cotterill 1968). In this context, the use of artificial Bragg structures could be envisaged so that this approach may be used with the soft X-ray radiation domain.

It is important to underline that the reflectance of the mirrors must be high. Indeed, the power P_n after the nth pass in a CBXFEL system is given by:

$$P_n = R\,(1+\mathcal{G})^n P_{n-1} + P_1 = \frac{[R\,(1+\mathcal{G})^n]-1}{R(1+\mathcal{G})}\,P_1 \qquad [8.1]$$

where R is the overall reflectance, the product of the reflectance of each mirror in the cavity and of the efficiency of any other optics that may be included in the chain, and $\mathcal{G}$ is the gain. From equation [8.1], it follows that the lasing threshold is given by:

$$R(1+\mathcal{G}) > 1 \qquad [8.2]$$

If we take into account the fact that the gain $\mathcal{G}$ varies as a function of the order of passage: $\mathcal{G} = \mathcal{G}(n)$, as shown in Figure 8.5, and in particular decreases significantly after a few dozen passes, it appears that the value of R must remain close to 1 so that the condition given by equation [8.2] is met, which remains a challenge.

6. See doi: 10.18429/JACoW-FEL2019-TUP033.

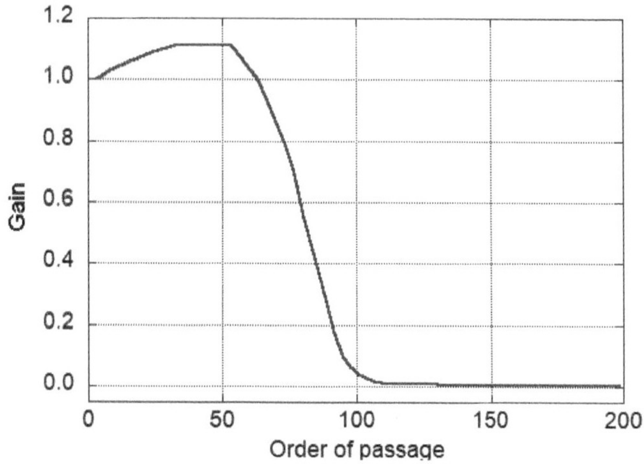

Figure 8.5. *Variation of the gain as a function of the order of passage in a CBXFEL-type system*

In the same way as for crystals in the hard X-ray domain, another challenge lies in the resistance to damage that is likely to be inflicted by short and intense radiation beams. The question of the behaviors of these optics in the temporal domain when they operate with short pulses (domain of a femtosecond) must also be examined in greater depth given that a body of work (see Chapter 3) has already highlighted modifications to their response compared to the case of a plane wave.

The defects listed above, inherent to the first generation of XFELs that use long undulators to achieve saturation and do not use self-seeding, have already been highlighted in an article by Piestrup (1988). Seeking to remedy these problems, he proposed in that article to charge a resonant cavity with a periodic dielectric structure likely to generate resonant transition radiation (RTR). The argument is that spontaneous emission of RTR is intense with respect to its yield per electron, if it is compared, for example, with the phenomenon of synchrotron emission, which makes RTR an excellent candidate for achieving a stimulated regime and obtaining lasing. The experimental results concerning X-RTR, presented in Chapter 7, support this approach. We could envisage using a resonant cavity based on artificial Bragg structures to perform retroaction of an X-RTR laser, as schematized in Figure 8.6 (Reid and Piestrup 1991). For the soft X-ray radiation domain, there is a complementary approach to the so-called SASE-XFELO approach previously mentioned.

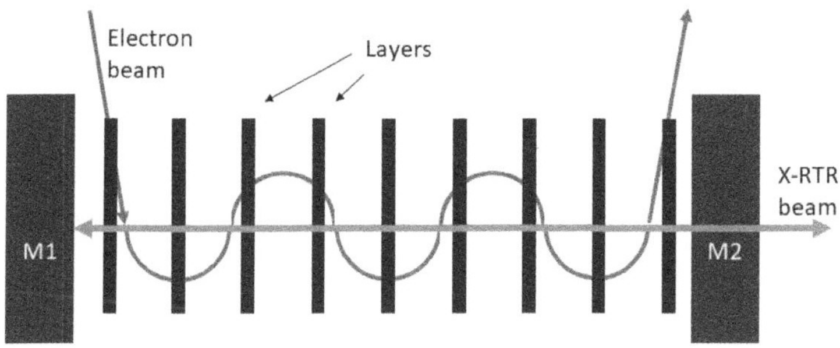

Figure 8.6. *Scheme of an X-ray laser using X-ray-stimulated resonant transition radiation (XRTR), using an amplifying cavity made of mirrors M1 and M2*

Even though the trend for improving the performances of an X-ray laser is to return to a cavity composed of reflectors *external* to the lasing medium, work starting in 1977 by A. Yariv regarding the feasibility of the laser has demonstrated that this approach required in all cases considerable pumping power (Yariv 1977). As an alternative, the use of a distributed feedback laser (DFBL) involving Bragg diffraction of the radiation stimulated within a spatially periodic medium has been proposed (Fisher 1974a, 1974b; Yariv 1974). A multilayer periodic structure seems to be a good candidate for creating this medium in order to achieve a laser in the soft X-ray radiation and EUV domain (Yariv 1977). It appears that the conditions are effectively met for this type of distributed laser to be successful: indeed, a series of experiments has just demonstrated that it is possible to obtain stimulated X-ray fluorescence radiation from a *solid target pumped by an XFEL* (Beye et al. 2013; Yoneda et al. 2015; Jonnard et al. 2017), and recently, stimulated emission in a solid has also been performed in elastic scattering (resonant elastic X-ray scattering (REXS)) and inelastic scattering (resonant inelastic X-ray scattering (RIXS)) regimes (Higley et al. 2022).

In addition, it has been demonstrated experimentally that certain multilayer Bragg structures resist the incidence of pumping XFEL for quite a long time, maintaining a periodic structure likely to result in retroaction (Chapman et al. 2006). Using the elements collected during studies of stimulated X-ray emission in backscattering mode, it was possible to fine-tune a feasibility study concerning X-ray distributed feedback lasers (X-DFBL) based on a multilayer structure operating in this mode: it has been demonstrated (Beye et al. 2013; Jonnard et al. 2017) that, for a solid homogeneous medium, the stimulated emission operated in a prioritized direction resulting from a compromise between the radiation penetration length from pumping L_i (XFEL) and the escape depth L_S of the stimulated radiation. Figure 8.6 shows the scheme of an X-DFBL operating in reflection mode and pumping performed by an XFEL. It is important to carefully choose the geometry and characteristics (type and thicknesses of layers, etc.) of the Bragg structure being used as a resonant cavity, to optimize the production of stimulated emission at a given photon energy (André et al. 2014). Therefore, an area of optimization consists of approaching the angles θ_2 and θ_1 (see Figure 8.7). The lasing conditions can be obtained in an initial approach using the coupled-wave theory (André et al. 2014).

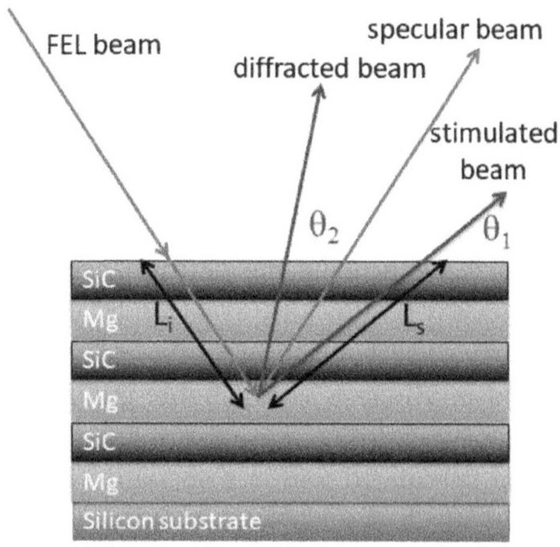

Figure 8.7. *Scheme of an X-ray distributed feedback laser consisting of a Bragg structure amplifying the emission of stimulated X-ray fluorescence pumped by an XFEL; the direction of diffracted radiation and the stimulated prioritized direction of the radiation are identified respectively by the angles θ_2 and θ_1 measured with respect to the plane of stratification. According to André et al. (2014)*[7]*. For a color version of this figure, see www.iste.co.uk/andre/xrayradiation.zip*

Physically, X-DFBLs are considered to use the Kossel effect (see Chapter 7) in the stimulated regime, in other words the emission of X-ray fluorescence stimulated and diffracted by the Bragg structure. This phenomenon was studied in detail by O. Peyrusse et al., in particular in the context of the Bloch–Maxwell model, for the case of pumping performed by XFELs (Peyrusse et al. 2017, 2020). Some experiments attempted to show an X-DFBL effect but were unsuccessful. It appears that the failure is related to the fact that the density of pumping energy available in these attempts is still insufficient, as corroborated by the calculations made with the models available (Peyrusse et al. 2017, 2020).

7. See doi: 10.1088/1054-660X/24/8/085001.

Analog experiments will very likely be performed with configurations that allow an increase in this density or a relaxation of the requirement for it. To do so, focusing of the pumping radiation could be used, using optics with artificial Bragg structures in the EUV domain and also from the point of view of relaxation, but also with the technique of seeding that could allow the intensity of pumping to be reduced by a factor of 100, as recent works on seeded stimulated emission in the X-ray radiation domain have led us to believe (Doyle et al. 2023). The proof of principle could also be obtained in the hard X-ray radiation domain, but using natural crystals.

8.3. References

Als-Nielsen, J. and McMorrow, D. (2000). *Elements of Modern X-ray Physics*. John Wiley & Sons, Hoboken.

André, J. M. (2023). *Cristaux et quasi-cristaux photoniques unidimensionnels : principe et théorie*. Auto-édition JMA, Amboise.

André, J.-M., Le Guen, K., Jonnard, P. (2014). Feasibility considerations of a soft-X-ray distributed feedback laser pumped by an X-ray free electron laser. *Laser Physics*, 24(8), 085001.

André, J.-M., Jonnard, P., Peyrusse, O. (2023). Self-seeding of SASE-soft-X-ray FEL using artificial Bragg reflectors [Online]. Available at: https://hal.science/hal-04157803 [Accessed 10 July 2023].

Aquila, A., Drescher, M., Laarmann, T., Barthelmeß, M., Chapman, H.N., Bajt, S. (2011). Moving the frontier of quantum control into the soft X-ray spectrum. *International Journal of Optics*, e417075.

Attwood, D. (1999). *Soft X-Rays and Extreme Ultraviolet Radiation: Principles and Applications*. Cambridge University Press, Cambridge.

Bachelard, R., Mercère, P., Idir, M., Couprie, M.-E., Labat, M., Chubar, O., Lambert, G., Zeitoun, P., Kimura, H., Ohashi, H. et al. (2011). Wavefront analysis of nonlinear self-amplified spontaneous-emission free-electron laser harmonics in the single-shot regime. *Physical Review Letters*, 106(23), 234801.

Beye, M., Schreck, S., Sorgenfrei, F., Trabant, C., Pontius, N., Schüßler-Langeheine, C., Wurth, W., Föhlisch, A. (2013). Stimulated X-ray emission for materials science. *Nature*, 501(7466), 191–194.

Chapman, H.N. (1996). Phase-retrieval X-ray microscopy by Wigner-distribution deconvolution. *Ultramicroscopy*, 66(3), 153–172.

Chapman, H.N., Barty, A., Bogan, M.J., Boutet, S., Frank, M., Hau-Riege, S.P., Marchesini, S., Woods, B.W., Bajt, S., Benner, W.H. et al. (2006). Femtosecond diffractive imaging with a soft-X-ray free-electron laser. *Nature Physics*, 2(12), 839–843.

Cotterill, R.M.J. (1968). A universal planar X-ray resonator. *Applied Physics Letters*, 12(12), 403–404.

Deacon, D.A.G., Elias, L.R., Madey, J.M.J., Ramian, G.J., Schwettman, H.A., Smith, T.I. (1977). First operation of a free-electron laser. *Physical Review Letters*, 38(16), 892–894.

Doyle, M.D., Halavanau, A., Zhang, Y., Michine, Y., Everts, J., Fuller, F., Alonso-Mori, R., Yabashi, M., Inoue, I., Osaka, T. et al. (2023). Seeded stimulated X-ray emission at 5.9 keV. *Optica*, 10(4), 513.

Dresselhaus, J.L., Fleckenstein, H., Domaracký, M., Prasciolu, M., Ivanov, N., Carnis, J., Murray, K.T., Morgan, A.J., Chapman, H.N., Bajt, S. (2022). Precise wavefront characterization of X-ray optical elements using a laboratory source. *Review of Scientific Instruments*, 93(7), 073704.

Duguay, M.A. and Rentzepis, P.M. (1967). Some approaches to vacuum UV and X-ray lasers. *Applied Physics Letters*, 10(12), 350–352.

Erko, A., Idir, M., Krist, T., Michette, A.G. (eds) (2008). *Modern Developments in X-Ray and Neutron Optics*. Springer, Berlin/Heidelberg [Online]. Available at: http://link.springer.com/10.1007/978-3-540-74561-7 [Accessed 20 June 2023].

Fisher, R.A. (1974a). On the possibility of a distributed feedback X-ray laser. *IEEE Journal of Quantum Electronics*, 10(9), 779–780.

Fisher, R.A. (1974b). Possibility of a distributed-feedback X-ray laser. *Applied Physics Letters*, 24(12), 598–599.

Furman, S.A. and Tikhonravov, A.V. (1992). *Basics of Optics of Multilayer Systems*. Editions Frontières, Gif-sur-Yvette.

Guérineau, N., Harchaoui, B., Primot, J. (2000). Talbot experiment re-examined: Demonstration of an achromatic and continuous self-imaging regime. *Optics Communications*, 180(4), 199–203.

Halavanau, A., Benediktovitch, A., Lutman, A.A., DePonte, D., Cocco, D., Rohringer, N., Bergmann, U., Pellegrini, C. (2020). Population inversion X-ray laser oscillator. *Proceedings of the National Academy of Sciences*, 117(27), 15511–15516.

Haner, M. and Warren, W.S. (1988). Synthesis of crafted optical pulses by time domain modulation in a fiber-grating compressor. *Applied Physics Letters*, 52(18), 1458–1460.

Hang, S., Tang, X., Li, H., Liu, Y., Mu, J., Zhou, W. (2019). Novel approach for Mars entry blackout elimination based on X-ray communication. *Journal of Spacecraft and Rockets*, 56(5), 1546–1552.

Higley, D.J., Chen, Z., Beye, M., Hantschmann, M., Reid, A.H., Mehta, V., Hellwig, O., Dakovski, G.L., Mitra, A., Engel, R.Y. et al. (2022). Stimulated resonant inelastic X-ray scattering in a solid. *Communications Physics*, 5(1), 83.

Hruszkewycz, S.O., Allain, M., Holt, M.V., Murray, C.E., Holt, J.R., Fuoss, P.H., Chamard, V. (2017). High-resolution three-dimensional structural microscopy by single-angle Bragg ptychography. *Nature Materials*, 16(2), 244–251.

Huang, Z. and Ruth, R.D. (2006). Fully coherent X-ray pulses from a regenerative-amplifier free-electron laser. *Physical Review Letters*, 96(14), 144801.

Idir, M., Cocco, D., Huang, L. (2022). Special issue "EUV and X ray wavefront sensing". *Sensors*, 22(10), 3940.

Joannopoulos, J.D., Johnson, S.G., Winn, J.N., Meade, R.D. (2011). *Photonic Crystals: Molding the Flow of Light*. Princeton University Press, Princeton [Online]. Available at: https://collaborate.princeton.edu/en/publications/photonic-crystals-molding-the-flow-of-light [Accessed 20 June 2023].

Jonnard, P., André, J.-M., Le Guen, K., Wu, M., Principi, E., Simoncig, A., Gessini, A., Mincigrucci, R., Masciovecchio, C., Peyrusse, O. (2017). EUV stimulated emission from MgO pumped by FEL pulses. *Structural Dynamics*, 4(5), 054306.

Kim, K.-J. and Shvyd'ko, Y.V. (2009). Tunable optical cavity for an X-ray free-electron-laser oscillator. *Physical Review Special Topics – Accelerators and Beams*, 12(3), 030703.

Kim, K.-J., Shvyd'ko, Y.V., Reiche, S. (2008). A proposal for an X-ray free-electron laser oscillator with an energy-recovery Linac. *Physical Review Letters*, 100(24), 244802.

Krzywiński, J., Feng, Y., Halavanau, A., Huang, Z., Kiss, A., MacArthur, J., Marcus, G., Sato, T., Zhu, D. (2019). Q-switching of X-ray optical cavities by using boron doped buried layer under a surface of a diamond crystal. In *39th Free Electron Laser Conference (FEL'19)*. JACOW Publishing, Geneva/Hamburg, 122–125 [Online]. Available at: https://accelconf.web.cern.ch/fel2019/doi/JACoW-FEL2019-TUP033.html [Accessed 16 May 2023].

Le Pape, S., Zeitoun, P., Hergot, J.-F., Carré, B., Dhez, P., François, M., Idir, M., Merdji, H., Ros, D., Carillon, A. et al. (2001). Measurement of the wavefront of XUV sources. *Le Journal de Physique IV*, 11(PR2), 533–536.

Liss, K.-D., Hock, R., Gomm, M., Waibel, B., Magerl, A., Krisch, M., Tucoulou, R. (2000). Storage of X-ray photons in a crystal resonator. *Nature*, 404(6776), 371–373.

Litvinenko, V.N., Burnham, B., Madey, J.M.J., Wu, Y. (1995). Dynamics of the Duke storage ring UV FEL. *Nuclear Instruments and Methods in Physics Research Section A: Accelerators, Spectrometers, Detectors and Associated Equipment*, 358(1), 369–373.

Madey, J.M.J. (1971). Stimulated emission of Bremsstrahlung in a periodic magnetic field. *Journal of Applied Physics*, 42(5), 1906–1913.

Mercère, P., Idir, M., Zeitoun, P., Levecq, X., Dovillaire, G., Bucourt, S., Douillet, D., Goldberg, K.A., Naulleau, P.P., Rekawa, S. (2004). X-ray wavefront Hartmann sensor. *AIP Conference Proceedings*, 705(1), 819–822.

Michette, A.G. (1986). *Optical Systems for Soft X Rays*. Springer, New York.

Nag, A., Chaphekar, P.A., Goswami, D. (2010). Applying genetic algorithm optimization to a folded geometry acousto-optic modulated spatial pulse shaper. *Review of Scientific Instruments*, 81(1), 013101.

Nematollahi, M., Lucke, P., Bayraktar, M., Yakshin, A., Rijnders, G., Bijkerk, F. (2019). Nanoscale piezoelectric surface modulation for adaptive extreme ultraviolet and soft X-ray optics. *Optics Letters*, 44(20), 5104–5107.

Pepper, D.M., Fekete, D., Yariv, A. (2008). Observation of amplified phase-conjugate reflection and optical parametric oscillation by degenerate four-wave mixing in a transparent medium. *Applied Physics Letters*, 33(1), 41–44.

Peyrusse, O., André, J.-M., Jonnard, P., Gaudin, J. (2017). Modeling of the interaction of an X-ray free-electron laser with large finite samples. *Physical Review E*, 96(4), 043205.

Peyrusse, O., Jonnard, P., Le Guen, K., André, J.-M. (2020). X-ray emission from layered media irradiated by an X-ray free-electron laser. *Physical Review A*, 101(1), 013818.

Piestrup, M. (1988). An X-ray free-electron laser loaded with a periodic dielectric. *IEEE Journal of Quantum Electronics*, 24(4), 591–597.

Rauer, P., Decking, W., Lipka, D., Thoden, D., Wohlenberg, T., Bahns, I., Brueggmann, U., Casalbuoni, S., Di Felice, M., Dommach, M. et al. (2023). Cavity-based X-ray free electron laser demonstrator at the European X-ray free electron laser facility. *Physical Review Accelerators and Beams*, 26(2), 020701.

Reid, M.B. and Piestrup, M.A. (1991). Resonance transition radiation X-ray lasers. *IEEE Journal of Quantum Electronics*, 27(11), 2440–2455.

Rizzi, J., Weitkamp, T., Guérineau, N., Idir, M., Mercère, P., Druart, G., Vincent, G., Silva, P. da, Primot, J. (2011). Quadriwave lateral shearing interferometry in an achromatic and continuously self-imaging regime for future X-ray phase imaging. *Optics Letters*, 36(8), 1398–1400.

Schawlow, A.L. and Townes, C.H. (1958). Infrared and optical masers. *Physical Review*, 112(6), 1940–1949.

Shvyd'ko, Yu.V., Lerche, M., Wille, H.-C., Gerdau, E., Lucht, M., Rüter, H.D., Alp, E.E., Khachatryan, R. (2003). X-ray interferometry with microelectronvolt resolution. *Physical Review Letters*, 90(1), 013904.

Shvyd'ko, Y.V., Stoupin, S., Cunsolo, A., Said, A.H., Huang, X. (2010). High-reflectivity high-resolution X-ray crystal optics with diamonds. *Nature Physics*, 6(3), 196–199.

Sutter, J.P., Chater, P.A., Signorato, R., Keeble, D.S., Hillman, M.R., Tucker, M.G., Alcock, S.G., Nistea, I., Wilhelm, H. (2019). A novel, 1 m long multilayer-coated piezo deformable bimorph mirror for focusing high-energy X-rays. *AIP Conference Proceedings*, 2054(1), 030005.

Uzoma, P.C., Shabbir, S., Hu, H., Okonkwo, P.C., Penkov, O.V. (2021). Multilayer reflective coatings for BEUV lithography: A review. *Nanomaterials*, 11(11), 2782.

Wang, Y., Liu, Y., Mu, J., Feng, Z., Tang, X. (2021). Collimating/focusing optical system designed for hard X-ray communication. *Nuclear Instruments and Methods in Physics Research Section A: Accelerators, Spectrometers, Detectors and Associated Equipment*, 1016, 165776.

Weiner, A.M., Heritage, J.P., Kirschner, E.M. (1988). High-resolution femtosecond pulse shaping. *JOSA B*, 5(8), 1563–1572.

Yariv, A. (1974). Analytical considerations of Bragg coupling coefficients and distributed-feedback X-ray lasers in single crystals. *Applied Physics Letters*, 25(2), 105–107.

Yariv, A. (1977). Some feasibility considerations of X-ray laser action. *Optics Communications*, 22(1), 1–4.

Yariv, A. and Pepper, D.M. (1977). Amplified reflection, phase conjugation, and oscillation in degenerate four-wave mixing. *Optics Letters*, 1(1), 16–18.

Yeh, P. (2005). *Optical Waves in Layered Media*. John Wiley & Sons, New York [Online]. Available at: http://eu.wiley.com/WileyCDA/WileyTitle/productCd-047 1731927.html [Accessed 16 April 2013].

Yoneda, H., Inubushi, Y., Nagamine, K., Michine, Y., Ohashi, H., Yumoto, H., Yamauchi, K., Mimura, H., Kitamura, H., Katayama, T. et al. (2015). Atomic inner-shell laser at 1.5-ångström wavelength pumped by an X-ray free-electron laser. *Nature*, 524(7566), 446–449.

8.4. In more detail

We indicate here some books in which certain points discussed in this book are developed in further detail.

Als-Nielsen, J. and McMorrow, D. (2000). *Elements of Modern X-ray Physics Applications*, 2nd edition. Wiley, New York.

André, J.-M. (2023). *Cristaux et quasi-cristaux unidimensionnels : principe et théorie*. Auto-édition JMA, Amboise.

Attwoood, D. and Sakdinawat, A. (1999). *X-rays and Extreme Ultraviolet Radiation: Principles and Applications*. Cambridge University Press, Cambridge.

Erko, A., Idir, M., Krist, T., Michette, A.G. (2008). *Modern Developments in X-ray and Neutron Optics*. Springer, Berlin.

Furman, S.A. and Tikhonravov, A.V. (1992). *Basics of Optics of Multilayer Systems*. Editions Frontières, Gif-sur-Yvette.

Joannopoulous, J.D., Johnson, S.G., Winn, J.N., Meade, R.D. (2001). *Photonics Crystals: Molding the Flow of Light*, 2nd edition. Princeton University Press, Princeton.

Michette, A.G. (1986). *Optical Systems for Soft X-rays*. Plenum Publishers, New York.

Yeh, P. (2005). Optical waves in layered media. In *Pure and Applied Optics*, Goodman, J.W. (ed.). John Wiley & Sons, New York.

Appendices

Appendix 1

Optimization of the Reflectance of a Bilayer System

To obtain the equation for reflectance optimization of a bilayer multilayer Bragg mirror, in terms of ratio γ and period d, we consider the reflection amplitude ρ_∞ in the case where the number of bilayers N_{bi} becomes very large (typically $N_{bi} \rightarrow \infty$).

To do this, we start from the matrix formalism in the representation $\begin{pmatrix} \mathcal{F}(z) \\ \mathcal{R}(z) \end{pmatrix}$ (see Chapter 2). The incident field at the surface $z = 0$ is written using the matrix $\widehat{\wp}_{bi}$ of the unit cell (in this case, a bilayer) and of the field at the surface $z = L$, where L is the total thickness of the structure:

$$\begin{pmatrix} \mathcal{F}(0) \\ \mathcal{R}(0) \end{pmatrix} = (\widehat{\wp}_{bi})^{N_{bi}} \begin{pmatrix} \mathcal{F}(L) \\ \mathcal{R}(L) \end{pmatrix} \quad [A1.1]$$

with:

$$\widehat{\wp}_{bi} = \frac{1}{1-r^2} \begin{pmatrix} exp(i\,\phi) - r^2 exp(-i\,\Delta) & r\big(exp(-i\,\phi) - exp(i\,\Delta)\big) \\ r\big(exp(i\,\phi) - exp(-i\,\Delta)\big) & exp(-i\,\phi) - r^2 exp(i\,\Delta) \end{pmatrix} \quad [A1.2]$$

where r is the Fresnel coefficient associated with the reflection at the interface of the layers i and $i+1$ and ϕ_i is the phase shift associated with traversing the layer i where:

$$\phi = \phi_1 + \phi_2, \Delta = \phi_1 - \phi_2 \quad [A1.3]$$

X-ray Radiation and Artificial Bragg Structures,
coordinated by Jean-Michel ANDRÉ and Philippe JONNARD. © ISTE Ltd 2025.

Expressing $\begin{pmatrix}\mathcal{F}(L)\\ \mathcal{R}(L)\end{pmatrix}$ using the eigenvectors $\boldsymbol{Z}^+ = \begin{pmatrix}V^+\\ W^+\end{pmatrix}$ and $\boldsymbol{Z}^- = \begin{pmatrix}V^-\\ W^-\end{pmatrix}$ of $\widehat{\wp}_{bi}$, it follows that:

$$\begin{pmatrix}\mathcal{F}(0)\\ \mathcal{R}(0)\end{pmatrix} = (\widehat{\wp}_{bi})^{N_{bi}}\,(a^+\,\boldsymbol{Z}^+ + a^-\,\boldsymbol{Z}^-) \qquad [A1.4]$$

so by introducing the eigenvalues P^+ of P^- of the eigensystem of $\widehat{\wp}_{bi}$:

$$\begin{pmatrix}\mathcal{F}(0)\\ \mathcal{R}(0)\end{pmatrix} = a^+\,(P^+)^{N_{bi}}\,\boldsymbol{Z}^+ + a^-(P^-)^{N_{bi}}\,\boldsymbol{Z}^- \qquad [A1.5]$$

Since $|P^-| < 1$, then:

$$\frac{lim}{N_{bi}\to\infty}\begin{pmatrix}\mathcal{F}(0)\\ \mathcal{R}(0)\end{pmatrix} = \frac{lim}{N_{bi}\to\infty}\,a^+\,(P^+)^{N_{bi}}\,\boldsymbol{Z}^+ \qquad [A1.6]$$

If we define the eigencoefficients of Fresnel reflection $\rho^{\pm}$ by:

$$\rho^{\pm} \equiv \frac{V^{\pm}}{W^{\pm}} \qquad [A1.7]$$

it follows, with explicit expressions for $V^{\pm}$ and $W^{\pm}$ using the elements ${\wp_{bi}}^{lk}$ of $\widehat{\wp}_{bi}$, that:

$$\rho^{\pm} = \frac{P^{\pm} - {\wp_{bi}}^{11}}{{\wp_{bi}}^{12}} \qquad [A1.8]$$

Using explicit expressions for ${\wp_{bi}}^{11}$ and ${\wp_{bi}}^{12}$ (see Chapter 2), we obtain:

$$\rho^{\pm} = exp(i\,\phi_2)\left(\mathcal{L} \pm \sqrt{\mathcal{L}^2 - 1}\right) \qquad [A1.9]$$

with:

$$\mathcal{L} = \frac{sin(\phi)}{2\,r\,sin(\phi_1)} \qquad [A1.10]$$

From equation [A1.6], it follows that:

$$\rho_\infty \equiv \frac{\mathcal{R}_\infty(0)}{\mathcal{F}_\infty(0)} = \frac{V^+}{W^+} \equiv \rho^+ \qquad [A1.11]$$

We obtain the optimization condition from:

$$\frac{d\rho_\infty}{d\phi_1} \approx \frac{d\mathcal{L}}{d\varphi_1} = 0\,,\varphi_1 = Re(\phi_1) \qquad [A1.12]$$

which, after derivation, and taking into account that $\mathcal{L}$ varies as a function of $\frac{1}{sin(\varphi_1)}(\beta_2\pi - (\beta_2 - \beta_1)\varphi_1)$, leads to the optimization equation:

$$tan(\varphi_1) = \varphi_1 - \frac{\beta_2}{\beta_2 - \beta_1}\pi \qquad [A1.13]$$

In the first-order Bragg condition, equation [A1.13] is reduced to:

$$tan(\pi\gamma) = \pi\gamma - \frac{\beta_2}{\beta_2 - \beta_1}\pi \qquad [A1.14]$$

Appendix 2

Polarization: Stokes Vector

To analyze the polarization of electromagnetic radiation, its electric field is separated into orthogonal components s and p which are respectively perpendicular (transverse electric (TE)) and orthogonal (transverse magnetic (TM)) to the plane of incidence. These components are written as:

$$E_{s,p} = \mathcal{E}_{s,p}\, e^{i(\omega\, t - k\, z + \psi_{s,p})} \quad \text{[A2.1]}$$

in an orthogonal identification system (x, y, z), where **z** is the axis of propagation. The polarization is described using a quantity known as the Stokes vector $\boldsymbol{S} = (S_0, S_1, S_2, S_3)$, where the quantities S_i, $i = 1, \ldots, 4$ are Stokes parameters.

Component $S_0 = \langle (\mathcal{E}_s)^2 + (\mathcal{E}_p)^2 \rangle$ corresponds to the intensity, component $S_1 = \langle (\mathcal{E}_p)^2 - (\mathcal{E}_s)^2 \rangle$ corresponds to the linear polarization along the **x** or **y** axis, component $S_2 = \langle 2\, \mathcal{E}_p \mathcal{E}_s\, cos(\psi_p - \psi_s) \rangle$ corresponds to the linear polarization along an axis turned by $\pm 45°$ with respect to the **x** axis, and component $S_3 = \langle 2\, \mathcal{E}_p \mathcal{E}_s\, sin(\psi_s - \psi_p) \rangle$ corresponds to the linear polarization along an axis turned by 135° with respect to the **x** axis and provides the degree of circular polarization. The symbol $\langle \ldots \rangle$ denotes an averaging operation with respect to the time. S_0 is normalized to a unit and the other parameters are included in the interval [–1,+1].

The Stokes parameters present the advantage of being usable for *partially* polarized radiation and they verify the following inequality, where the term DOP denotes the degree of polarization:

X-ray Radiation and Artificial Bragg Structures,
coordinated by Jean-Michel ANDRÉ and Philippe JONNARD. © ISTE Ltd 2025.

$$1 \geq DOP^2 \equiv \frac{S_1{}^2+S_2{}^2+S_3{}^2}{S_0{}^2} \qquad [A2.2]$$

For non-polarized radiation $S_1 = S_2 = S_3 = 0$, whereas for totally polarized radiation $S_1{}^2 + S_2{}^2 + S_3{}^2 = 1$. In practice, the polarization is determined using a phase retarder (mirror or plate), which performs a phase shift (phase delay) between the component *p* and the component *s*, and an analyzer (mirror), which preferentially reflects one of these components. The measurement is made by performing a rotation of these optical elements with respect to the optical axis **z** (see Figure A2.1).

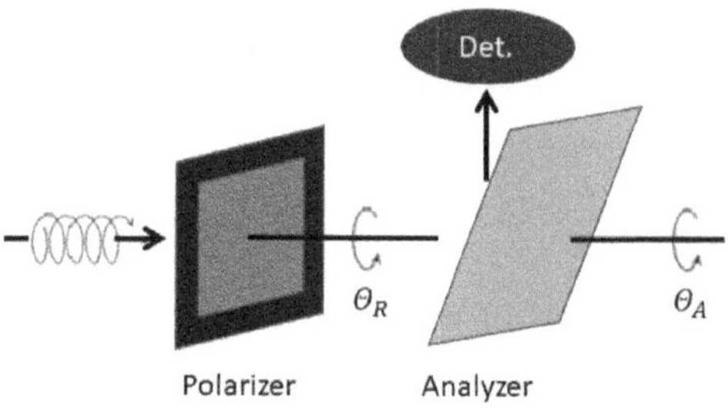

Figure A2.1. *Scheme of a polarization measurement; "Det." is the detector*

From a mathematical point of view, the action of an optical element on the state of polarization of the radiation is described by a matrix known as the Müller matrix *M*, which is of the form:

$$M(\Psi,\Delta) = \begin{pmatrix} 1 & -\cos(2\Psi) & 0 & 0 \\ -\cos(2\Psi) & 1 & 0 & 0 \\ 0 & 0 & \sin(2\Psi)\cos(\Delta) & \sin(2\Psi)\sin(\Delta) \\ 0 & 0 & -\sin(2\Psi)\sin(\Delta) & \sin(2\Psi)\cos(\Delta) \end{pmatrix} \qquad [A2.3]$$

where Δ is the phase delay and Ψ is the phase of the ratio of the transmittance (reflectance) of the component *p* to the transmittance (reflectance) of the component *s*, for a transmission retarder (reflection). The expression of the Müller matrix depends on its orientation with regard to the incident beam. If the optical element is turned with respect to the latter at an angle Θ without a change in the *angle of incidence*, the initial Müller matrix will undergo a transformation associated with the rotation matrix $R(\Theta)$:

$$R(\Theta) = \begin{pmatrix} 1 & 0 & 0 & 0 \\ 0 & cos(2\Theta) & sin(2\Theta) & 0 \\ 0 & -sin(2\Theta) & cos(2\Theta) & 0 \\ 0 & 0 & 0 & 1 \end{pmatrix} \qquad [A2.4]$$

Consequently, the final Müller matrix obtained after a rotation by an angle Θ_R at the retarder and a rotation of an angle Θ_A at the analyzer is given by:

$$M(\Psi_f, \Delta_f) = W(\Theta_A, \Psi_A, \Delta_A, \Theta_R, \Psi_R, \Delta_R)\, M(\Psi_i, \Delta_i)$$

$$W(\Theta_A, \Psi_A, \Delta_A, \Theta_R, \Psi_R, \Delta_R) = R(-\Theta_A)\, M(\Psi_A, \Delta_A)\, R(\Theta_A)\, R(-\Theta_R)\, M(\Psi_R, \Delta_R)\, R(\Theta_R) \qquad [A2.5]$$

The indices *A*, *R*, *f* and *i* correspond respectively to the first letter of the terms "Analyzer", "Retarder", "final" and "initial" (incident radiation to analyze).

The final Stokes vector $\boldsymbol{S_f}$ is related to the initial Stokes vector $\boldsymbol{S_i}$ by:

$$\boldsymbol{S_f} = W(\Theta_A, \Psi_A, \Delta_A, \Theta_R, \Psi_R, \Delta_R)\, \boldsymbol{S_i} \qquad [A2.6]$$

The quantity measured by the detector is the intensity, in other words the parameter 0 of $\boldsymbol{S_f}$ (S_{f0} not given explicitly). It contains the four parameters of the initial Stokes vector $\boldsymbol{S_i}$ and the phase delays Δ, phases Ψ and angles of rotation $\Theta_A, \Psi_A, \Delta_A, \Theta_R, \Psi_R$ and Δ_R. In fact, Δ_A is only involved if the detector is sensitive to polarization of the detected radiation. A mathematical procedure (e.g. an adjustment using the least squares method), using the expression for S_{f0} and the measured value of this quantity, enables other parameters of the initial Stokes vector to be reached by varying Θ_A and Θ_R and consequently determination of the state of polarization of the incident radiation.

List of Authors

Jean-Michel ANDRÉ
LCPMR, CNRS
Sorbonne Université
Paris
France

Philippe JONNARD
LCPMR, CNRS
Sorbonne Université
Paris
France

Karine LE GUEN
LCPMR, CNRS
Sorbonne Université
Paris
France

Index

A, B

C, D

E, F

H, I

L, M

O, P

R, S

T, U, V